RUST FOR RUSTACEANS

DAVID ROBINSON
A0 61 363 25274
June 20 2022

RUST FOR RUSTACEANS

Idiomatic Programming for Experienced Developers

by Jon Gjengset

no starch
press

San Francisco

Printed in the United States of America

Second printing

25 24 23 22   2 3 4 5 6 7 8 9

ISBN-13: 978-1-7185-0185-0 (print)
ISBN-13: 978-1-7185-0186-7 (ebook)

Publisher: William Pollock
Managing Editor: Jill Franklin
Production Manager and Editor: Rachel Monaghan
Developmental Editor: Liz Chadwick
Cover Illustrator: James L. Barry
Interior Design: Octopod Studios
Technical Reviewer: David Tolnay
Copyeditor: Rachel Head
Compositor: Maureen Forys, Happenstance Type-O-Rama
Proofreader: Sadie Barry
Indexer: Beth Nauman-Montana

For information on book distributors or translations, please contact No Starch Press, Inc. directly:
No Starch Press, Inc.
245 8th Street, San Francisco, CA 94103
phone: 1.415.863.9900; info@nostarch.com
www.nostarch.com

Library of Congress Control Number: 2021944983

About the Author

Jon Gjengset has worked in the Rust ecosystem since the early days of Rust 1.0, and built a high-performance relational database from scratch in Rust over the course of his PhD at MIT. He's been a frequent contributor to the Rust toolchain and ecosystem, including the asynchronous runtime Tokio, and maintains several popular Rust crates, such as hdrhistogram and inferno. Jon has been teaching Rust since 2018, when he started live-streaming intermediate-level Rust programming sessions. Since then, he's made videos that cover advanced topics like async and await, pinning, variance, atomics, dynamic dispatch, and more, which have been received enthusiastically by the Rust community.

About the Technical Reviewer

David Tolnay is a prolific, well-known, and respected contributor in the Rust ecosystem who maintains some of the most widely used Rust libraries, including syn, serde, and anyhow. He is also a member of the Rust library team.

BRIEF CONTENTS

CONTENTS IN DETAIL

3
DESIGNING INTERFACES
37

4
ERROR HANDLING
57

5
PROJECT STRUCTURE
67

FOREWORD

Dear reader,

In the course of your experience with Rust so far, it's likely that you have noticed a knowledge gap between what your existing learning resources have prepared you for versus what you see from the folks making the top tier of widely used Rust libraries and applications.

Libraries that do very well are commonly powered by a co-occurrence of taste and dedication on the authors' part: feeling what to build, and building the thing (it's that simple). This book teaches neither of those things.

However, it's been my experience that taste emerges from a deep comfort with the fundamental pieces. It's here I feel this book will be helpful to you. I don't consider it a coincidence that pretty much all of the "household name" open source Rust library developers understand everything in this book—even when it's not the case that they *use* every single thing from the book in every single library.

In this book you will find a level of nuance and tradeoffs and opinions that does not arise from introductory material. Structs are structs, and we have no need to have an opinion about structs. But infinitely flexible macro APIs (Chapter 7), the judicious application of unsafe code (Chapter 9), effective testing that speeds you up rather than slowing you down (Chapter 6)—someone who's digested *The Rust Programming Language* (a.k.a. The Book, *https://doc.rust-lang.org/book/*) but not much beyond that is generally going to have a hard time manifesting what they know into high-quality or innovative projects, but this book takes you to the *starting point* to begin building your personal taste in highly polished Rust development. You will take what you read here and get it wrong a bunch of times, and get it right a couple, and get better.

I encourage you to seize upon that starting point consciously. I want you to be free to think that we got something wrong in this book; that the best current guidance in here is missing something, and that you can accomplish something over the next couple years that is better than what anybody else has envisioned. That's how Rust and its ecosystem have gotten to this point.

David Tolnay

PREFACE

One of the goals listed on the Rust 2018 roadmap was to develop teaching resources to better serve intermediate Rustaceans—those who aren't beginners but also aren't compiler experts looking to design a new iteration of the borrow checker. That call inspired me to start live-streaming coding sessions where I implemented real systems in Rust in real time—not toy projects or long-winded introductions to basic concepts, but libraries and tools I would actually use for my research. My thinking was that Rust newcomers needed to see an experienced Rust programmer go through the whole development process, including design, debugging, and iteration, in order to understand how to *think in Rust*. While a beginner could attempt the same things themselves, it'd likely be far slower and frustrating since they would also be learning the language along the way.

Many developers said that my videos provided a good way to learn to use Rust "for real," which was very exciting. However, over the years, it also became clear that the videos weren't for everyone, or for every situation. Some developers prefer to be more in control over their own learning and would rather have a teaching resource they can consume at their own pace. Others just need to understand a particular topic better, or find out how a specific feature works or is best used, and for those situations, a six-hour coding video isn't that helpful. I wanted to make sure that intermediate resources were available for those people and situations as well, which is what ultimately made me decide to write this book. My aim was to distill all that time spent teaching intermediate Rust by example into solid textual explanations of the most important intermediate topics.

I realized early on that the book would complement the videos, not replace them. I remain convinced that the best way to quickly gain experience in a language, barring actively working with it yourself daily, is to watch someone experienced use it. But in my time writing this book, I've also found that this format works incredibly well as a comprehensive, by-topic reference that collects lots of knowledge in one place, which is where coding videos fall terribly short. The coding sessions help develop your Rust experience, intuition, and taste. The book teaches you the theory, mechanisms, and idioms of the language. And ultimately, a developer needs all of the above to truly excel at what they do.

Now, many many words and iterations later, what you have in front of you is my attempt at plugging another hole in the set of intermediate Rust teaching resources. I hope that you find it useful and that we're now one step closer to fulfilling that roadmap goal!

ACKNOWLEDGMENTS

Having never written a book before, I knew little of what to expect from the process. I naively assumed that it would be like writing a sequence of blog posts, or perhaps like writing extensive documentation, but writing a book has been a whole different ballgame. I've spent countless hours researching, planning, writing, rewriting, discarding, rethinking, and editing. And I'm incredibly grateful for the support and patience my girlfriend, Talia, has shown throughout the many late nights I spent working on this project— without you, the writing experience would have been so much bleaker.

This book could never have happened without the incredible Rust developers and community, who have developed a language and ecosystem I continue to find a joy to interact with and am inspired to help spread and teach to the best of my ability. The same goes for the amazing people who have watched my streams over the years; I don't think I would ever have ended up in a position to write this book in the first place without your ongoing support, encouragement, and endless curiosity.

This book also would not have been half as good had it not been for David Tolnay, who to my great delight agreed to be the book's technical reviewer. David was obviously invaluable in finding errors in both theory and code, but it was his vast experience, attention to detail, and penchant for pedagogy that truly made a mark. His thoughtful and insightful comments sometimes made me decide to rewrite entire sections, but always in ways that made them immeasurably better than they had been before.

The same goes for my editor, Liz Chadwick, and the rest of the publishing team at No Starch. It was so fun to see Liz's journey through the book's development; she picked up Rust along the way while reading, and I was thrilled when her comments showed that she truly followed along with the intermediate material. The discussions whenever there was something she didn't follow were always illuminating, and resulted in more accessible and thorough explanations.

I'd also be remiss not to mention Steve Klabnik and Carol Nichols, the authors of *The Rust Programming Language*, which was my first introduction to Rust. This book is, at least in my mind, very much a sequel to their book and could not exist without the extraordinary job they've done of making the fundamentals of Rust so easily accessible and well explained.

Finally, I want to give a nod to you. Yes, you! Writing this book has been a very long process, and it's partially the outpouring of support and encouragement from the people wanting to read it that has kept me going throughout it all. And it's people like you who pick up this book, whether virtually or physically, with a desire to improve your own understanding and skills that drive me to keep contributing to the collection of Rust teaching resources as best I can.

Thank you all!

INTRODUCTION

In any language, the gap between what the introductory material teaches you and what you know after years of hands-on experience is always wide. Over time, you build familiarity with idioms, develop better mental models for core concepts, learn which designs and patterns work and which do not, and discover useful libraries and tools in the surrounding ecosystem. Taken together, this experience enables you to write better code in less time.

With this book, I'm hoping to distill years of my own experience writing Rust code into a single, easy-to-digest resource. *Rust for Rustaceans* picks up where *The Rust Programming Language* ("the Rust book") leaves off, though it's well suited to any Rust programmer that wants to go beyond the basics, wherever you learned the trade. This book delves deeper into concepts such as unsafe code, the trait system, no_std code, and macros. It also covers new areas like asynchronous I/O, testing, embedded development, and ergonomic API design. I aim to explain and demystify these more advanced and powerful features of Rust and to enable you to build faster, more ergonomic, and more robust applications going forward.

What's in the Book

This book is written both as a guide and as a reference. The chapters are more or less independent, so you can skip directly to topics that particularly interest you (or are currently causing you headaches), or you can read the book start to finish for a more holistic experience. That said, I do recommend that you start by reading Chapters 1 and 2, as they lay the foundation for the later chapters and for many topics that will come up in your day-to-day Rust development. Here's a quick breakdown of what you'll find in each chapter:

Chapter 1, *Foundations*, gives deeper, more thorough descriptions of fundamental Rust concepts like variables, memory, ownership, borrowing, and lifetimes that you'll need to be familiar with to follow the remainder of the book.

Chapter 2, *Types*, similarly provides a more exhaustive explanation of types and traits in Rust, including how the compiler reasons about them, their features and restrictions, and a number of advanced applications.

Chapter 3, *Designing Interfaces*, covers how to design APIs that are intuitive, flexible, and misuse-resistant, including advice on how to name things, how to use the type system to enforce API contracts, and when to use generics versus trait objects.

Chapter 4, *Error Handling*, explores the two primary kinds of errors (enumerated and opaque), when the use of each is appropriate, and how each of these are defined, constructed, propagated, and handled.

Chapter 5, *Project Structure*, focuses on the non-code parts of a Rust project, such as Cargo metadata and configuration, crate features, and versioning.

Chapter 6, *Testing*, details how the standard Rust testing harness works and presents some testing tools and techniques that go beyond standard unit and integration tests, such as fuzzing and performance testing.

Chapter 7, *Macros*, covers both declarative and procedural macros, including how they're written, what they're useful for, and some of their pitfalls.

Chapter 8, *Asynchronous Programming*, gives an introduction to the difference between synchronous and asynchronous interfaces and then delves into how asynchrony is represented in Rust both at the low level of Future and Pin and at the high level of async and await. The chapter also explains the role of an asynchronous executor and how it makes the whole async machinery come together.

Chapter 9, *Unsafe Code*, explains the great powers that the unsafe keyword unlocks and the great responsibilities that come with those powers. You'll learn about common gotchas in unsafe code as well as tools and techniques you can use to reduce the risk of incorrect unsafe code.

Chapter 10, *Concurrency (and Parallelism)*, looks at how concurrency is represented in Rust and why it can be so difficult to get right in terms of both correctness and performance. It covers how concurrency and asynchrony are related (but not the same), how concurrency works when you get closer to the hardware, and how to stay sane while trying to write correct concurrent programs.

Chapter 11, *Foreign Function Interfaces*, teaches you how to make Rust cooperate nicely with other languages and what FFI primitives like the extern keyword actually do.

Chapter 12, *Rust Without the Standard Library*, is all about using Rust in situations where the full standard library isn't available, such as on embedded devices or other constrained platforms, where you're restricted to what the core and alloc modules provide.

Chapter 13, *The Rust Ecosystem*, doesn't cover a particular Rust subject but instead aims to give broader guidance about working in the Rust ecosystem. It contains descriptions of common design patterns, advice on staying up to date on additions to the language and best practices, tips on useful tools and other useful trivia I've accumulated over the years that isn't otherwise described in any single place.

The book has a website at *https://rust-for-rustaceans.com* with links to resources from the book, future errata, and the like. You'll also find that information at the book's page on the No Starch Press website at *https://nostarch.com/rust-rustaceans/*.

And now, with all that out of the way, there's only one thing left to do:

```
fn main() {
```

1

FOUNDATIONS

As you dive into the more advanced corners of Rust, it's important that you ensure you have a solid understanding of the fundamentals. In Rust, as in any programming language, the precise meaning of various keywords and concepts becomes important as you begin to use the language in more sophisticated ways. In this chapter, we'll walk through many of Rust's primitives and try to define more clearly what they mean, how they work, and why they are exactly the way that they are. Specifically, we'll look at how variables and values differ, how they are represented in memory, and the different memory regions a program has. We'll then discuss some of the subtleties of ownership, borrowing, and lifetimes that you'll need to have a handle on before you continue with the book.

You can read this chapter from top to bottom if you wish, or you can use it as a reference to brush up on the concepts that you feel less sure about. I recommend that you move on only when you feel completely comfortable with the content of this chapter, as misconceptions about how these primitives work will quickly get in the way of understanding the more advanced topics, or lead to you using them incorrectly.

Talking About Memory

Not all memory is created equal. In most programming environments, your programs have access to a stack, a heap, registers, text segments, memory-mapped registers, memory-mapped files, and perhaps nonvolatile RAM. Which one you choose to use in a particular situation has implications for what you can store there, how long it remains accessible, and what mechanisms you use to access it. The exact details of these memory regions vary between platforms and are beyond the scope of this book, but some are so important to how you reason about Rust code that they are worth covering here.

Memory Terminology

Before we dive into regions of memory, you first need to know about the difference between values, variables, and pointers. A *value* in Rust is the combination of a type and an element of that type's domain of values. A value can be turned into a sequence of bytes using its type's *representation*, but on its own you can think of a value more like what you, the programmer, meant. For example, the number 6 in the type u8 is an instance of the mathematical integer 6, and its in-memory representation is the byte 0x06. Similarly, the str "Hello world" is a value in the domain of all strings whose representation is its UTF-8 encoding. A value's meaning is independent of the location where those bytes are stored.

A value is stored in a *place*, which is the Rust terminology for "a location that can hold a value." This place can be on the stack, on the heap, or in a number of other locations. The most common place to store a value is a *variable*, which is a named value slot on the stack.

A *pointer* is a value that holds the address of a region of memory, so the pointer points to a place. A pointer can be dereferenced to access the value stored in the memory location it points to. We can store the same pointer in more than one variable and therefore have multiple variables that indirectly refer to the same location in memory and thus the same underlying value.

Consider the code in Listing 1-1, which illustrates these three elements.

```
let x = 42;
let y = 43;
let var1 = &x;
let mut var2 = &x;
❶ var2 = &y;
```

Listing 1-1: Values, variables, and pointers

Here, there are four distinct values: 42 (an i32), 43 (an i32), the address of x (a pointer), and the address of y (a pointer). There are also four variables: x, y, var1, and var2. The latter two variables both hold values of the pointer type, because references are pointers. While var1 and var2 store the same value initially, they store separate, independent copies of that value; when we change the value stored in var2 ❶, the value in var1 does not change. In particular, the = operator stores the value of the right-hand side expression in the place named by the left-hand side.

An interesting example of where the distinction between variables, values, and pointers becomes important is in a statement such as:

```
let string = "Hello world";
```

Even though we assign a string value to the variable string, the *actual* value of the variable is a pointer to the first character in the string value "Hello world", and not the string value itself. At this point you might say, "But hang on, where is the string value stored, then? Where does the pointer point?" If so, you have a keen eye—we'll get to that in a second.

NOTE *Technically, the value of string also includes the string's length. We'll talk about that in Chapter 2 when we discuss wide pointer types.*

Variables in Depth

The definition of a variable I gave earlier is broad and unlikely to be all that useful in and of itself. As you encounter more complex code, you'll need a more accurate mental model to help you reason through what the programs are really doing. There are many such models that we can make use of. Describing them all in detail would take up several chapters and is beyond the scope of this book, but broadly speaking, they can be divided into two categories: high-level models and low-level models. High-level models are useful when thinking about code at the level of lifetimes and borrows, while low-level models are good for when you are reasoning about unsafe code and raw pointers. The models for variables described in the following two sections will suffice for most of the material in this book.

High-Level Model

In the high-level model, we don't think of variables as places that hold bytes. Instead, we think of them just as names given to values as they are instantiated, moved, and used throughout a program. When you assign a value to a variable, that value is from then on named by that variable. When a variable is later accessed, you can imagine drawing a line from the previous access of that variable to the new access, which establishes a dependency relationship between the two accesses. If the value in a variable is moved, no lines can be drawn from it anymore.

In this model, a variable exists only so long as it holds a legal value; you cannot draw lines from a variable whose value is uninitialized or has been moved, so effectively it isn't there. Using this model, your entire program consists of many of these dependency lines, often called *flows*, each one tracing the lifetime of a particular instance of a value. Flows can fork and merge when there are branches, with each split tracing a distinct lifetime for that value. The compiler can check that at any given point in your program, all flows that can exist in parallel with each other are compatible. For example, there cannot be two parallel flows with mutable access to a value. Nor can there be a flow that borrows a value while there is no flow that owns the value. Listing 1-2 shows examples of both of these cases.

```
let mut x;
// this access would be illegal, nowhere to draw the flow from:
// assert_eq!(x, 42);
❶ x = 42;
// this is okay, can draw a flow from the value assigned above:
❷ let y = &x;
// this establishes a second, mutable flow from x:
❸ x = 43;
// this continues the flow from y, which in turn draws from x.
// but that flow conflicts with the assignment to x!
❹ assert_eq!(*y, 42);
```

Listing 1-2: Illegal flows that the borrow checker will catch

First, we cannot use x before it is initialized, because we have nowhere to draw the flow from. Only when we assign a value to x can we draw flows from it. This code has two flows: one exclusive (&mut) flow from ❶ to ❸, and one shared (&) flow from ❶ through ❷ to ❹. The borrow checker inspects every vertex of every flow and checks that no other incompatible flows exist concurrently. In this case, when the borrow checker inspects the exclusive flow at ❸, it sees the shared flow that terminates at ❹. Since you cannot have an exclusive and a shared use of a value at the same time, the borrow checker (correctly) rejects the code. Notice that if ❹ was not there, this code would compile fine! The shared flow would terminate at ❷, and when the exclusive flow is checked at ❸, no conflicting flows would exist.

If a new variable is declared with the same name as a previous one, they are still considered distinct variables. This is called *shadowing*—the later variable "shadows" the former by the same name. The two variables coexist, though subsequent code no longer has a way to name the earlier one. This model matches roughly how the compiler, and the borrow checker in particular, reasons about your program, and is actually used internally in the compiler to produce efficient code.

Low-Level Model

Variables name memory locations that may or may not hold legal values. You can think of a variable as a "value slot." When you assign to it, the slot is filled, and its old value (if it had one) is dropped and replaced. When you access it, the compiler checks that the slot isn't empty, as that would mean the variable is uninitialized or its value has been moved. A pointer to a variable refers to the variable's backing memory and can be dereferenced to get at its value. For example, in the statement let x: usize, the variable x is a name for a region of memory on the stack that has room for a value the size of a usize, though it does not have a well-defined value (its slot is empty)..If you assign a value to that variable, such as with x = 6, that region of memory will then hold the bits representing the value 6. &x does not change when you assign to x. If you declare multiple variables with the same name, they still end up with different chunks of memory backing them. This model matches the memory model used by C and C++, and many other low-level languages, and is useful for when you need to reason explicitly about memory.

NOTE *In this example, we ignore CPU registers and treat them as an optimization. In reality, the compiler may use a register to back a variable instead of a region of memory if no memory address is needed for that variable.*

You may find that one of these matches your previous model better than the other, but I urge you to try to wrap your head around both of them. They are both equally valid, and both are simplifications, like any useful mental model has to be. If you are able to consider a piece of code from both of these perspectives, you will find it much easier to work through complicated code segments and understand why they do or do not compile and work as you expect.

Memory Regions

Now that you have a grip on how we refer to memory, we need to talk about what memory actually is. There are many different regions of memory, and perhaps surprisingly, not all of them are stored in the DRAM of your computer. Which part of memory you use has a significant impact on how you write your code. The three most important regions for the purposes of writing Rust code are the stack, the heap, and static memory.

The Stack

The *stack* is a segment of memory that your program uses as scratch space for function calls. Each time a function is called, a contiguous chunk of memory called a *frame* is allocated at the top of the stack. Near the bottom of the stack is the frame for the main function, and as functions call other functions, additional frames are pushed onto the stack. A function's frame contains all the variables within that function, along with any arguments the function takes. When the function returns, its stack frame is reclaimed.

The bytes that make up the values of the function's local variables are not immediately wiped, but it's not safe to access them as they may have been overwritten by a subsequent function call whose frame overlaps with the reclaimed one. And even if they haven't been overwritten, they may contain values that are illegal to use, such as ones that were moved when the function returned.

Stack frames, and crucially the fact that they eventually disappear, are very closely tied to the notion of lifetimes in Rust. Any variable stored in a frame on the stack cannot be accessed after that frame goes away, so any reference to it must have a lifetime that is at most as long as the lifetime of the frame.

The Heap

The *heap* is a pool of memory that isn't tied to the current call stack of the program. Values in heap memory live until they are explicitly deallocated. This is useful when you want a value to live beyond the lifetime of the current function's frame. If that value is the function's return value, the calling function can leave some space on its stack for the called function to write that value into before it returns. But if you want to, say, send that value to a different thread with which the current thread may share no stack frames at all, you can store it on the heap.

The heap allows you to explicitly allocate contiguous segments of memory. When you do so, you get a pointer to the start of that segment of memory. That memory segment is reserved for you until you later deallocate it; this process is often referred to as *freeing*, after the name of the corresponding function in the C standard library. Since allocations from the heap do not go away when a function returns, you can allocate memory for a value in one place, pass the pointer to it to another thread, and have that thread safely continue to operate on that value. Or, phrased differently, when you heap-allocate memory, the resulting pointer has an unconstrained lifetime—its lifetime is however long your program keeps it alive.

The primary mechanism for interacting with the heap in Rust is the Box type. When you write Box::new(value), the value is placed on the heap, and what you are given back (the Box<T>) is a pointer to that value on the heap. When the Box is eventually dropped, that memory is freed.

If you forget to deallocate heap memory, it will stick around forever, and your application will eventually eat up all the memory on your machine. This is called *leaking memory* and is usually something you want to avoid. However, there are some cases where you explicitly want to leak memory. For example, say you have a read-only configuration that the entire program should be able to access. You can allocate that on the heap and explicitly leak it with Box::leak to get a 'static reference to it.

Static Memory

Static memory is really a catch-all term for several closely related regions located in the file your program is compiled into. These regions are automatically loaded into your program's memory when that program is executed. Values in static memory live for the entire execution of your program. Your program's static memory contains the program's binary code, which is usually mapped as read-only. As your program executes, it walks through the binary code in the text segment instruction by instruction and jumps around whenever a function is called. Static memory also holds the memory for variables you declare with the static keyword, as well as certain constant values in your code, like strings.

The special lifetime 'static, which gets its name from the static memory region, marks a reference as being valid for "as long as static memory is around," which is until the program shuts down. Since a static variable's memory is allocated when the program starts, a reference to a variable in static memory is, by definition, 'static, as it is not deallocated until the program shuts down. The inverse is not true—there can be 'static references that do not point to static memory—but the name is still appropriate: once you create a reference with a static lifetime, whatever it points to might as well be in static memory as far as the rest of the program is concerned, as it can be used for however long your program wishes.

You will encounter the 'static lifetime much more often than you will encounter truly static memory (through the static keyword, for example) when working with Rust. This is because 'static often shows up in trait bounds on type parameters. A bound like T: 'static indicates that the type parameter T is able to live for however long we keep it around for, up to and including the remaining execution of the program. Essentially, this bound requires that T is owned and self-sufficient, either in that it does not borrow other (non-static) values or that anything it does borrow is also 'static and thus will stick around until the end of the program. A good example of 'static as a bound is the std::thread::spawn function that creates a new thread, which requires that the closure you pass it is 'static. Since the new thread may outlive the current thread, the new thread cannot refer to anything stored on the old thread's stack. The new thread can refer only to values that will live for its entire lifetime, which may be for the remaining duration of the program.

NOTE *You may wonder how const differs from static. The const keyword declares the following item as constant. Constant items can be completely computed at compile time, and any code that refers to them is replaced with the constant's computed value during compilation. A constant has no memory or other storage associated with it (it is not a place). You can think of constant as a convenient name for a particular value.*

Ownership

Rust's memory model centers on the idea that all values have a single *owner*—that is, exactly one location (usually a scope) is responsible for ultimately deallocating each value. This is enforced through the borrow checker. If the value is moved, such as by assigning it to a new variable, pushing it to a vector, or placing it on the heap, the ownership of the value moves from the old location to the new one. At that point, you can no longer access the value through variables that flow from the original owner, even though the bits that make up the value are technically still there. Instead, you must access the moved value through variables that refer to its new location.

Some types are rebels and do not follow this rule. If a value's type implements the special Copy trait, the value is not considered to have moved even if it is reassigned to a new memory location. Instead, the value is *copied*, and both the old and new locations remain accessible. Essentially, another identical instance of that same value is constructed at the destination of the move. Most primitive types in Rust, such as the integer and floating-point types, are Copy. To be Copy, it must be possible to duplicate the type's values simply by copying their bits. This eliminates all types that *contain* non-Copy types as well as any type that owns a resource it must deallocate when the value is dropped.

To see why, consider what would happen if a type like Box were Copy. If we executed box2 = box1, then box1 and box2 would both believe that they owned the heap memory allocated for the box, and they would both attempt to free it when they went out of scope. Freeing the memory twice could have catastrophic consequences.

When a value's owner no longer has use for it, it is the owner's responsibility to do any necessary cleanup for that value by *dropping* it. In Rust, dropping happens automatically when the variable that holds the value is no longer in scope. Types usually recursively drop values they contain, so dropping a variable of a complex type may result in many values being dropped. Because of Rust's discrete ownership requirement, we cannot accidentally drop the same value multiple times. A variable that holds a reference to another value does not own that other value, so the value isn't dropped when the variable drops.

The code in Listing 1-3 gives a quick summary of the rules around ownership, move and copy semantics, and dropping.

```
let x1 = 42;
let y1 = Box::new(84);
{ // starts a new scope
❶ let z = (x1, y1);
  // z goes out of scope, and is dropped;
  // it in turn drops the values from x1 and y1
❷ }
  // x1's value is Copy, so it was not moved into z
❸ let x2 = x1;
  // y1's value is not Copy, so it was moved into z
❹ // let y2 = y1;
```

Listing 1-3: Moving and copying semantics

We start out with two values, the number 42 and a Box (a heap-allocated value) containing the number 84. The former is Copy, whereas the latter is not. When we place x1 and y1 into the tuple z ❶, x1 is *copied* into z, whereas y1 is *moved* into z. At this point, x1 continues to be accessible and can be used again ❸. On the other hand, y1 is rendered inaccessible once its value has been moved ❹, and any attempt to access it would incur a compiler error. When z goes out of scope ❷, the tuple value it contains is dropped, and this in turn drops the value copied from x1 and the one moved from y1. When the Box from y1 is dropped, it also deallocates the heap memory used to store y1's value.

Rust automatically drops values when they go out of scope, such as x1 and y1 in the inner scope in Listing 1-3. The rules for the order in which to drop are fairly simple: variables (including function arguments) are dropped in reverse order, and nested values are dropped in source-code order.

This might sound weird at first—why the discrepancy? If we look at it closely, though, it makes a lot of sense. Say you write a function that declares a string and then inserts a reference to that string into a new hash table. When the function returns, the hash table must be dropped first; if the string were dropped first, the hash table would then hold an invalid reference! In general, later variables may contain references to earlier values, whereas the inverse cannot happen due to Rust's lifetime rules. And for that reason, Rust drops variables in reverse order.

Now, we could have the same behavior for nested values, like the values in a tuple, array, or struct, but that would likely surprise users. If you constructed an array that contained two values, it'd seem odd if the last element of the array were dropped first. The same applies to tuples and structs, where the most intuitive behavior is for the first tuple element or field to be dropped first, then the second, and so on. Unlike for variables, there is no need to reverse the drop order in this case, since Rust doesn't (currently) allow self-references in a single value. So, Rust goes with the intuitive option.

Borrowing and Lifetimes

Rust allows the owner of a value to lend out that value to others, without giving up ownership, through references. *References* are pointers that come with an additional contract for how they can be used, such as whether the reference provides exclusive access to the referenced value, or whether the referenced value may also have other references point to it.

Shared References

A shared reference, &T, is, as the name implies, a pointer that may be shared. Any number of other references may exist to the same value, and each shared reference is Copy, so you can trivially make more of them. Values behind shared references are not mutable; you cannot modify or reassign the value a shared reference points to, nor can you cast a shared reference to a mutable one.

The Rust compiler is allowed to assume that the value a shared reference points to *will not change* while that reference lives. For example, if the Rust compiler sees that the value behind a shared reference is read multiple times in a function, it is within its rights to read it only once and reuse that value. More concretely, the assertion in Listing 1-4 should never fail.

```
fn cache(input: &i32, sum: &mut i32) {
  *sum = *input + *input;
  assert_eq!(*sum, 2 * *input);
}
```

Listing 1-4: Rust assumes that shared references are immutable.

Whether or not the compiler chooses to apply a given optimization is more or less irrelevant. The compiler heuristics change over time, so you generally want to code against what the compiler is allowed to do rather than what it actually does in a particular case at a particular moment in time.

Mutable References

The alternative to a shared reference is a mutable reference: &mut T. With mutable references, the Rust compiler is again allowed to make full use of the contract that the reference comes with: the compiler assumes that there are no other threads accessing the target value, whether through a shared reference or a mutable one. In other words, it assumes that the mutable reference is *exclusive*. This enables some interesting optimizations that are not readily available in other languages. Take, for example, the code in Listing 1-5.

```
fn noalias(input: &i32, output: &mut i32) {
  if *input == 1 {
  ❶ *output = 2;
  }
❷ if *input != 1 {
    *output = 3;
  }
}
```

Listing 1-5: Rust assumes that mutable references are exclusive.

In Rust, the compiler can assume that input and output do not point to the same memory. Therefore, the reassignment of output at ❶ cannot affect the check at ❷, and the entire function can be compiled as a single if-else block. If the compiler could not rely on the exclusive mutability contract, that optimization would be invalid, since an input of 1 could then result in an output of 3 in a case like noalias(&x, &mut x).

A mutable reference lets you mutate only the memory location that the reference points to. Whether you can mutate values that lie beyond the immediate reference depends on the methods provided by the type that lies between. This may be easier to understand with an example, so consider Listing 1-6.

```
let x = 42;
let mut y = &x; // y is of type &i32
let z = &mut y; // z is of type &mut &i32
```

Listing 1-6: Mutability applies only to the immediately referenced memory.

In this example, you are able to change the value of the pointer y to a different value (that is, a different pointer) by making it reference a different variable, but you cannot change the value that is pointed to (that is, the value of x). Similarly, you can change the pointer value of y through z, but you cannot change z itself to hold a different reference.

The primary difference between owning a value and having a mutable reference to it is that the owner is responsible for dropping the value when it is no longer necessary. Apart from that, you can do anything through a mutable reference that you can if you own the value, with one caveat: if you move the value behind the mutable reference, then you must leave another value in its place. If you did not, the owner would still think it needed to drop the value, but there would be no value for it to drop!

Listing 1-7 gives an example of the ways in which you can move the value behind a mutable reference.

```
fn replace_with_84(s: &mut Box<i32>) {
  // this is not okay, as *s would be empty:
❶ // let was = *s;
  // but this is:
❷ let was = std::mem::take(s);
  // so is this:
❸ *s = was;
  // we can exchange values behind &mut:
  let mut r = Box::new(84);
❹ std::mem::swap(s, &mut r);
  assert_ne!(*r, 84);
}
let mut s = Box::new(42);
replace_with_84(&mut s);
❺
```

Listing 1-7: Access through a mutable reference must leave a value behind.

I've added commented-out lines that represent illegal operations. You cannot simply move the value out ❶ since the caller would still think they owned that value and would free it again at ❺, leading to a double free. If you just want to leave some valid value behind, std::mem::take ❷ is a good candidate. It is equivalent to std::mem::replace(&mut value, Default::default()); it moves value out from behind the mutable reference but leaves a new, default value for the type in its place. The default is a separate, owned value, so it is safe for the caller to drop it when the scope ends at ❺.

Alternatively, if you don't need the old value behind the reference, you can overwrite it with a value that you already own ❸, leaving it to the caller to drop the value later. When you do this, the value that used to be behind the mutable reference is dropped immediately.

Finally, if you have two mutable references, you can swap their values without owning either of them ❹, since both references will end up with a legal owned value for their owners to eventually free.

Interior Mutability

Some types provide *interior mutability*, meaning they allow you to mutate a value through a shared reference. These types usually rely on additional mechanisms (like atomic CPU instructions) or invariants to provide safe mutability without relying on the semantics of exclusive references. These normally fall into two categories: those that let you get a mutable reference through a shared reference, and those that let you replace a value given only a shared reference.

The first category consists of types like Mutex and RefCell, which contain safety mechanisms to ensure that, for any value they give a mutable reference to, only one mutable reference (and no shared references) can exist at a time. Under the hood, these types (and those like them) all rely on a type called UnsafeCell, whose name should immediately make you hesitate to use it. We will cover UnsafeCell in more detail in Chapter 9, but for now you should know that it is the *only* correct way to mutate through a shared reference.

Other categories of types that provide interior mutability are those that do not give out a mutable reference to the inner value but instead just give you methods for manipulating that value in place. The atomic integer types in std::sync::atomic and the std::cell::Cell type fall into this category. You cannot get a reference directly to the usize or i32 behind such a type, but you can read and replace its value at a given point in time.

NOTE *The Cell type in the standard library is an interesting example of safe interior mutability through invariants. It is not shareable across threads and never gives out a reference to the value contained in the Cell. Instead, the methods all either replace the value entirely or return a copy of the contained value. Since no references can exist to the inner value, it is always okay to move it. And since Cell isn't shareable across threads, the inner value will never be concurrently mutated even though mutation happens through a shared reference.*

Lifetimes

If you're reading this book, you're probably already familiar with the concept of lifetimes, likely through repeated notices from the compiler about lifetime rules violations. That level of understanding will serve you well for the majority of Rust code you will write, but as we dive deeper into the more complex parts of Rust, you will need a more rigorous mental model to work with.

Newer Rust developers are often taught to think of lifetimes as corresponding to scopes: a lifetime begins when you take a reference to some variable and ends when that variable is moved or goes out of scope. That's often correct, and usually useful, but the reality is a little more complex. A *lifetime* is really a name for a region of code that some reference must be valid for. While a lifetime will frequently coincide with a scope, it does not have to, as we will see later in this section.

Lifetimes and the Borrow Checker

At the heart of Rust lifetimes is the *borrow checker*. Whenever a reference with some lifetime 'a is used, the borrow checker checks that 'a is still *alive*. It does this by tracing the path back to where 'a starts—where the reference was taken—from the point of use and checking that there are no conflicting uses along that path. This ensures that the reference still points to a value that it is safe to access. This is similar to the high-level "data flow" mental model we discussed earlier in the chapter; the compiler checks that the flow of the reference we are accessing does not conflict with any other parallel flows.

Listing 1-8 shows a simple code example with lifetime annotations for the reference to x.

```
  let mut x = Box::new(42);
❶ let r = &x;            // 'a
  if rand() > 0.5 {
❷ *x = 84;
  } else {
❸ println!("{}", r);  // 'a
  }
❹
```

Listing 1-8: Lifetimes do not need to be contiguous.

The lifetime starts at ❶ when we take a reference to x. In the first branch ❷, we then immediately try to modify x by changing its value to 84, which requires a &mut x. The borrow checker takes out a mutable reference to x and immediately checks its use. It finds no conflicting uses between when the reference was taken and when it was used, so it accepts the code. This may come as a surprise if you are used to thinking about lifetimes as scopes, since r is still in scope at ❷ (it goes out of scope at ❹). But the borrow checker is smart enough to realize that r is never used later if this branch is taken, and therefore it is fine for x to be mutably accessed here. Or, phrased differently, the lifetime created at ❶ does not extend into this branch: there is no flow from r beyond ❷, and therefore there are no conflicting flows. The borrow checker then finds the use of r in the print statement at ❸. It walks the path back to ❶ and finds no conflicting uses (❷ is not on that path), so it accepts this use as well.

If we were to add another use of r at ❹ in Listing 1-8, the code would no longer compile. The lifetime 'a would then last from ❶ all the way until ❹ (the last use of r), and when the borrow checker checked our new use of r, it would discover a conflicting use at ❷.

Lifetimes can get quite convoluted. In Listing 1-9 you can see an example of a lifetime that has *holes*, where it's intermittently invalid between where it starts and where it ultimately ends.

```
  let mut x = Box::new(42);
❶ let mut z = &x;              // 'a
  for i in 0..100 {
❷ println!("{}", z);           // 'a
❸ x = Box::new(i);
❹ z = &x;                      // 'a
  }
  println!("{}", z);           // 'a
```

Listing 1-9: Lifetimes can have holes.

The lifetime starts at ❶ when we take a reference to x. We then move out of x at ❸, which ends the lifetime 'a because it is no longer valid. The borrow checker accepts this move by considering 'a ended at ❷, which leaves no conflicting flows from x at ❸. Then, we restart the lifetime by updating the reference in z ❹. Regardless of whether the code now loops back around to ❷ or continues to the final print statement, both of those uses now have a valid value to flow from, and there are no conflicting flows, so the borrow checker accepts the code!

Again, this aligns perfectly with the data-flow model of memory we discussed earlier. When x is moved, z stops existing. When we reassign z later, we are creating an entirely new variable that exists only from that point forward. It just so happens that that new variable is also named z. With that model in mind, this example is not weird.

NOTE   *The borrow checker is, and has to be, conservative. If it's unsure whether a borrow is valid, it rejects it, as the consequences of allowing an invalid borrow could be disastrous. The borrow checker keeps getting smarter, but there are times when it needs help to understand why a borrow is legal. This is part of why we have unsafe Rust.*

Generic Lifetimes

Occasionally you need to store references within your own types. Those references need to have a lifetime so that the borrow checker can check their validity when they are used in the various methods on that type. This is especially true if you want a method on your type to return a reference that outlives the reference to self.

Rust lets you make a type definition generic over one or more lifetimes, just as it allows you to make it generic over types. *The Rust Programming Language* by Steve Klabnik and Carol Nichols (No Starch Press, 2018) covers this topic in some detail, so I won't reiterate the basics here. But as you write more complex types of this nature, there are two subtleties around the interaction between such types and lifetimes that you should be aware of.

First, if your type also implements Drop, then dropping your type counts as a use of any lifetime or type your type is generic over. Essentially, when an instance of your type is dropped, the borrow checker will check that it's still legal to use any of your type's generic lifetimes before dropping it. This is necessary in case your drop code *does* use any of those references. If your type does not implement Drop, dropping the type does *not* count as a use, and users are free to ignore any references stored in your type as long as they do not use it anymore, like we saw in Listing 1-7. We'll talk more about these rules around dropping in Chapter 9.

Second, while a type can be generic over multiple lifetimes, making it so often only serves to unnecessarily complicate your type signature. Usually, a type being generic over a single lifetime is fine, and the compiler will use the shorter of the lifetimes for any references inserted into your type as that one lifetime. You should only really use multiple generic lifetime parameters if you have a type that contains multiple references, and its methods return references that should be tied to the lifetime of only *one* of those references.

Consider the type in Listing 1-10, which gives you an iterator over parts of a string separated by a particular other string.

```
struct StrSplit<'s, 'p> {
  delimiter: &'p str,
  document: &'s str,
}
impl<'s, 'p> Iterator for StrSplit<'s, 'p> {
  type Item = &'s str;
  fn next(&self) -> Option<Self::Item> {
    todo!()
  }
}
fn str_before(s: &str, c: char) -> Option<&str> {
  StrSplit { document: s, delimiter: &c.to_string() }.next()
}
```

Listing 1-10: A type that needs to be generic over multiple lifetimes

When you construct this type, you have to give the delimiter and document to search, both of which are references to string values. When you ask for the next string, you get a reference into the document. Consider what would happen if you used a single lifetime in this type. The values yielded by the iterator would be tied to the lifetime of the document *and* the delimiter. This would make str_before impossible to write: the return type would have a lifetime associated with a variable local to the function—the String produced by to_string—and the borrow checker would reject the code.

Lifetime Variance

Variance is a concept that programmers are often exposed to but rarely know the name of because it's mostly invisible. At a glance, variance describes what types are subtypes of other types and when a subtype can be used in place of a supertype (and vice versa). Broadly speaking, a type A is a subtype of another type B if A is at least as useful as B. Variance is the reason why, in Java, you can pass a Turtle to a function that accepts an Animal if Turtle is a subtype of Animal, or why, in Rust, you can pass a &'static str to a function that accepts a &'a str.

While variance usually hides out of sight, it comes up often enough that we need to have a working knowledge of it. Turtle is a subtype of Animal because a Turtle is more "useful" than some unspecified Animal—a Turtle can do anything an Animal can do, and likely more. Similarly, 'static is a subtype of 'a because a 'static lives at least as long as any 'a and so is more useful. Or, more generally, if 'b: 'a ('b outlives 'a), then 'b is a subtype of 'a. This is obviously not the formal definition, but it gets close enough to be of practical use.

All types have a variance, which defines what other similar types can be used in that type's place. There are three kinds of variance: covariant, invariant, and contravariant. A type is *covariant* if you can just use a subtype in place of the type. For example, if a variable is of type &'a T, you can provide a value of type &'static T to it, because &'a T is covariant in 'a. &'a T is also covariant in T, so you can pass a &Vec<&'static str> to a function that takes &Vec<&'a str>.

Some types are *invariant*, which means that you must provide exactly the given type. &mut T is an example of this—if a function takes a &mut Vec<&'a str>, you cannot pass it a &mut Vec<&'static str>. That is, &mut T is invariant in T. If you could, the function could put a short-lived string inside the Vec, which the caller would then continue using, thinking that it were a Vec<&'static str> and thus that the contained string were 'static! Any type that provides mutability is generally invariant for the same reason—for example, Cell<T> is invariant in T.

The last category, *contravariance*, comes up for function arguments. Function types are more useful if they're okay with their arguments being *less* useful. This is clearer if you contrast the variance of the argument types on their own with their variance when used as function arguments:

```
let x: &'static str; // more useful, lives longer
let x: &'a      str; // less useful, lives shorter

fn take_func1(&'static str) // stricter, so less useful
fn take_func2(&'a str)      // less strict, more useful
```

This flipped relationship indicates that Fn(T) is contravariant in T.

So why do you need to learn about variance when it comes to lifetimes? Variance becomes relevant when you consider how generic lifetime parameters interact with the borrow checker. Consider a type like the one shown in Listing 1-11, which uses multiple lifetimes in a single field.

```
struct MutStr<'a, 'b> {
  s: &'a mut &'b str
}
let mut s = "hello";
❶ *MutStr { s: &mut s }.s = "world";
println!("{}", s);
```

Listing 1-11: A type that needs to be generic over multiple lifetimes

At first glance, using two lifetimes here seems unnecessary—we have no methods that need to differentiate between a borrow of different parts of the structure, as we did with StrSplit in Listing 1-10. But if you replace the two lifetimes here with a single 'a, the code no longer compiles! And it's all because of variance.

The syntax at ❶ *may seem alien. It's equivalent to defining a variable x holding a MutStr and then writing *x.s = "world", except that there's no variable and so the MutStr is dropped immediately.*

At ❶, the compiler must determine what lifetime the lifetime parameter(s) should be set to. If there are two lifetimes, 'a is set to the to-be-determined lifetime of the borrow of s, and 'b is set to 'static since that's the lifetime of the provided string "hello". If there is just one lifetime 'a, the compiler infers that that lifetime must be 'static.

When we later attempt to access the string reference s through a shared reference to print it, the compiler tries to shorten the mutable borrow of s used by MutStr to allow the shared borrow of s.

In the two-lifetime case, 'a simply ends just before the println, and 'b stays the same. In the single-lifetime case, on the other hand, we run into issues. The compiler wants to shorten the borrow of s, but to do so, it would also have to shorten the borrow of the str. While &'static str can in general be shortened to any &'a str (&'a T is covariant in 'a), here it's behind a &mut T, which is invariant in T. Invariance requires that the relevant type is never replaced with a sub- or supertype, so the compiler's attempt to shorten the borrow fails, and it reports that the list is still mutably borrowed. Ouch!

Because of the reduced flexibility imposed by invariance, you want to ensure that your types remain covariant (or contravariant where appropriate) over as many of their generic parameters as possible. If that requires introducing additional lifetime arguments, you need to carefully weigh the cognitive cost of adding another parameter against the ergonomic cost of invariance.

Summary

The aim of this chapter has been to establish a solid, shared foundation that we can build on in the chapters to come. By now, I hope you feel that you have a firm grasp on Rust's memory and ownership model, and that those errors you may have gotten from the borrow checker seem less mysterious. You might have known bits and pieces of what we covered here already, but hopefully the chapter has given you a more holistic image of how it all fits together. In the next chapter, we will do something similar for types. We'll go over how types are represented in memory, see how generics and traits produce running code, and take a look at some of the special type and trait constructs Rust offers for more advanced use cases.

2

TYPES

Now that the fundamentals are out of the way, we'll look at Rust's type system. We'll skip past the basics covered in *The Rust Programming Language* and instead dive head-first into how different types are laid out in memory, the ins and outs of traits and trait bounds, existential types, and the rules for using types across crate boundaries.

Types in Memory

Every Rust value has a type. Types serve many purposes in Rust, as we'll see in this chapter, but one of their most fundamental roles is to tell you how to interpret bits of memory. For example, the sequence of bits 0b10111101 (written in hexadecimal notation as 0xBD) does not mean anything in and of itself until you assign it a type. When interpreted under the type u8, that sequence of bits is the number 189. When interpreted under the type i8, it is –67. When you define your own types, it's the compiler's job to determine where each part of the defined type goes in the in-memory representation for that type. Where does each field of your struct appear in the sequence of bits? Where is the discriminant for your enum stored? It's important to understand how this process works as you begin to write more advanced Rust code, because these details affect both the correctness and the performance of your code.

Alignment

Before we talk about how a type's in-memory representation is determined, we first need to discuss the notion of *alignment*, which dictates where the bytes for a type can be stored. Once a type's representation has been determined, you might think you can take any arbitrary memory location and interpret the bytes stored there as that type. While that is true in a theoretical sense, in practice the hardware also constrains where a given type can be placed. The most obvious example of this is that pointers point to *bytes*, not *bits*. If you placed a value of type T starting at bit 4 of your computer's memory, you would have no way to refer to its location; you can create a pointer pointing only to byte 0 or byte 1 (bit 8). For this reason, all values, no matter their type, must start at a byte boundary. We say that all values must be at least *byte-aligned*—they must be placed at an address that is a multiple of 8 bits.

Some values have more stringent alignment rules than just being byte-aligned. In the CPU and the memory system, memory is often accessed in blocks larger than a single byte. For example, on a 64-bit CPU, most values are accessed in chunks of 8 bytes (64 bits), with each operation starting at an 8-byte-aligned address. This is referred to as the CPU's *word size*. The CPU then uses some cleverness to handle reading and writing smaller values, or values that span the boundaries of these chunks.

Where possible, you want to ensure that the hardware can operate in its "native" alignment. To see why, consider what happens if you try to read an i64 that starts in the middle of an 8-byte block (that is, the pointer to it is not 8-byte-aligned). The hardware will have to do two reads—one from the second half of the first block to get to the start of the i64, and one from the first half of the second block to read the rest of the i64—and then splice the results together. This is not very efficient. Since the operation is spread across multiple accesses to the underlying memory, you may also end up with strange results if the memory you're reading from is concurrently written to by a different thread. You might read the first 4 bytes before the other thread's write has happened and the second 4 bytes after, resulting in a corrupted value.

Operations on data that is not aligned are referred to as *misaligned accesses* and can lead to poor performance and bad concurrency problems. For this reason, many CPU operations require, or strongly prefer, that their arguments are *naturally aligned*. A naturally aligned value is one whose alignment matches its size. So, for example, for an 8-byte load, the provided address would need to be 8-byte-aligned.

Since aligned accesses are generally faster and provide stronger consistency semantics, the compiler tries to take advantage of them where possible. It does this by giving every type an alignment that's computed based on the types that it contains. Built-in values are usually aligned to their size, so a u8 is byte-aligned, a u16 is 2-byte-aligned, a u32 is 4-byte-aligned, and a u64 is 8-byte-aligned. Complex types—types that contain other types—are typically assigned the largest alignment of any type they contain. For example, a type that contains a u8, a u16, and a u32 will be 4-byte-aligned because of the u32.

Layout

Now that you know about alignment, we can explore how the compiler decides on the in-memory representation, known as the *layout*, of a type. By default, as you'll see shortly, the Rust compiler gives very few guarantees about how it lays out types, which makes for a poor starting point for understanding the underlying principles. Luckily, Rust provides a repr attribute you can add to your type definitions to request a particular in-memory representation for that type. The most common one you will see, if you see one at all, is repr(C). As the name suggests, it lays out the type in a way that is compatible with how a C or C++ compiler would lay out the same type. This is helpful when writing Rust code that interfaces with other languages using the foreign-function interface, which we'll talk about in Chapter 11, as Rust will generate a layout that matches the expectations of the other language's compiler. Since the C layout is predictable and not subject to change, repr(C) is also useful in unsafe contexts if you're working with raw pointers into the type, or if you need to cast between two different types that you know have the same fields. And, of course, it is perfect for taking our first steps into layout algorithms.

NOTE *Another useful representation is repr(transparent), which can be used only on types with a single field and which guarantees that the layout of the outer type is exactly the same as that of the inner type. This comes in handy in combination with the "newtype" pattern, where you may want to operate on the in-memory representations of some struct A and struct NewA(A) as if they were the same. Without repr(transparent), the Rust compiler does not guarantee that they will have the same layout.*

So, let's look how the compiler would lay out a particular type with repr(C): the Foo type in Listing 2-1. How do you think the compiler would lay this out in memory?

```
#[repr(C)]
struct Foo {
  tiny: bool,
  normal: u32,
  small: u8,
  long: u64,
  short: u16,
}
```

Listing 2-1: Alignment affects layout.

First the compiler sees the field tiny, whose logical size is 1 bit (true or false). But since the CPU and memory operate in terms of bytes, tiny is given 1 byte in the in-memory representation. Next, normal is a 4-byte type, so we want it to be 4-byte-aligned. But even if Foo is aligned, the 1 byte we allocated to tiny is going to make normal miss its alignment. To rectify this, the compiler inserts 3 bytes of *padding*—bytes with an indeterminate value that are ignored in user code—into the in-memory representation between tiny and normal. No values go into the padding, but it does take up space.

For the next field, small, alignment is simple: it's a 1-byte value, and the current byte offset into the struct is $1 + 3 + 4 = 8$. This is already byte-aligned, so small can go immediately after normal. With long we have a problem again, though. We are now $1 + 3 + 4 + 1 = 9$ bytes into Foo. If Foo is aligned, then long is not 8-byte-aligned the way we want it to be, so we must insert another 7 bytes of padding to make long aligned again. This also conveniently ensures the 2-byte alignment we need for the last field, short, bringing the total to 26 bytes. Now that we've gone through all the fields, we also need to determine the alignment of Foo itself. The rule here is to use the largest alignment of any of Foo's fields, which will be 8 bytes because of long. So, to ensure that Foo remains aligned if placed in, say, an array, the compiler then adds a final 6 bytes of padding to make Foo's size a multiple of its alignment at 32 bytes.

Now we are ready to shed the C legacy and consider what would happen to the layout if we did not use repr(C) in Listing 2-1. One of the primary limitations of the C representation is that it requires that we place all fields in the same order that they appear in the original struct definition. The default Rust representation repr(Rust) removes that limitation, along with a couple of other lesser restrictions, such as deterministic field ordering for types that happen to have the same fields. That is, even two different types that share all the same fields, of the same type, in the same order, are not guaranteed to be laid out the same when using the default Rust layout!

Since we're now allowed to reorder the fields, we can place them in decreasing order of size. This means we no longer need the padding between Foo's fields; the fields themselves are used to achieve the necessary alignment! Foo is now just the size of its fields: only 16 bytes. This is one of the reasons why Rust by default does not give many guarantees about how a type is laid out in memory: by giving the compiler more leeway to rearrange things, we can produce more efficient code.

It turns out there's also a third way to lay out a type, and that is to tell the compiler that we do not want any padding between our fields. In doing so, we're saying that we are willing to take the performance hit of using misaligned accesses. The most common use case for this is when the impact of every additional byte of memory can be felt, such as if you have a lot of instances of the type, if you have very limited memory, or if you're sending the in-memory representation over a lower-bandwidth medium like a network connection. To opt in to this behavior, you can annotate your type with #[repr(packed)]. Keep in mind that this may lead to much slower code, and in extreme cases, this can cause your program to crash if you try to perform operations that the CPU supports only on aligned arguments.

Sometimes, you want to give a particular field or type a larger alignment than it technically requires. You can do that using the attribute #[repr(align(n))]. A common use case for this is to ensure that different values stored contiguously in memory (like in an array) end up in different cache lines on the CPU. That way, you avoid *false sharing*, which can cause huge performance degradations in concurrent programs. False sharing occurs when two different CPUs access different values that happen to share a cache line; while they can theoretically operate in parallel, they both end up contending to update the same single entry in the cache. We'll talk about concurrency in much greater detail in Chapter 10.

Complex Types

You might be curious about how the compiler represents other Rust types in memory. Here's a quick reference:

Tuple   Represented like a struct with fields of the same type as the tuple values in the same order.

Array   Represented as a contiguous sequence of the contained type with no padding between the elements.

Union   Layout is chosen independently for each variant. Alignment is the maximum across all the variants.

Enumeration   Same as union, but with one additional hidden shared field that stores the enum variant discriminant. The discriminant is the value the code uses to determine which of the enum variants a given value holds. The size of the discriminant field depends on the number of variants.

Dynamically Sized Types and Wide Pointers

You may have come across the marker trait Sized in various odd corners of the Rust documentation and in error messages. Usually, it comes up because the compiler wants you to provide a type that is Sized, but you (apparently) did not. Most types in Rust implement Sized automatically— that is, they have a size that's known at compile time—but two common types do not: trait objects and slices. If you have, for example, a dyn Iterator or a [u8], those do not have a well-defined size. Their size depends on some information that is known only when the program runs and not at compile time, which is why they are called *dynamically sized types (DSTs)*. Nobody knows ahead of time whether the dyn Iterator your function received is this 200-byte struct or that 8-byte struct. This presents a problem: often the compiler must know the size of something in order to produce valid code, such as how much space to allocate to a tuple of type (i32, dyn Iterator, [u8], i32) or what offset to use if your code tries to access the fourth field. But if the type isn't Sized, that information isn't available.

The compiler requires types to be Sized nearly everywhere. Struct fields, function arguments, return values, variable types, and array types must all be Sized. This restriction is so common that every single type bound you write includes T: Sized unless you explicitly opt out of it with T: ?Sized (the ? means "may not be"). But this is pretty unhelpful if you have a DST and want to do something with it, like if you really want your function to accept a trait object or a slice as an argument.

The way to bridge this gap between unsized and sized types is to place unsized types behind a *wide pointer* (also known as a *fat pointer*). A wide pointer is just like a normal pointer, but it includes an extra word-sized field that gives the additional information about that pointer that the compiler needs to generate reasonable code for working with the pointer. When you take a reference to a DST, the compiler automatically constructs a wide pointer for you. For a slice, the extra information is simply the length of the slice. For a trait object—well, we'll get to that later. And crucially, that wide pointer *is* Sized. Specifically, it is twice the size of a usize (the size of a word on the target platform): one usize for holding the pointer, and one usize for holding the extra information needed to "complete" the type.

NOTE *Box and Arc also support storing wide pointers, which is why they both support T: ?Sized.*

Traits and Trait Bounds

Traits are a key piece of Rust's type system—they are the glue that allows types to interoperate even though they don't know about each other at the time they are defined. *The Rust Programming Language* does a great job of covering how to define and use traits, so I won't go over that here. Instead, we're going to take a look at some of the more technical aspects of traits: how they're implemented, restrictions you have to adhere to, and some more esoteric uses of traits.

Compilation and Dispatch

By now, you've probably written a decent amount of generic code in Rust. You've used generic type parameters on types and methods, and maybe even a few trait bounds here and there. But have you ever wondered what actually happens to generic code when you compile it, or what happens when you call a trait method on a dyn Trait?

When you write a type or function that is generic over T, you're really telling the compiler to make a copy of that type or function for each type T. When you construct a Vec<i32> or a HashMap<String, bool>, the compiler essentially copy-pastes the generic type and all its implementation blocks and replaces all instances of each generic parameter with the concrete type you provided. It makes a full copy of the Vec type with every T replaced with i32, and a full copy of the HashMap type with every K replaced with String and every V with bool.

NOTE *In reality, the compiler does not actually do a full copy-paste. It copies only parts of the code that you use, so if you never call find on a Vec<i32>, the code for find won't be copied and compiled.*

The same thing applies to generic functions. Consider the code in Listing 2-2, which shows a generic method.

```
impl String {
  pub fn contains(&self, p: impl Pattern) -> bool {
    p.is_contained_in(self)
  }
}
```

Listing 2-2: A generic method using static dispatch

A copy of this method is made for every distinct pattern type (recall that impl Trait is shorthand for <T: Trait>). We need a different copy of the function body for each impl Pattern type because we need to know the address of the is_contained_in function to call it. The CPU needs to be told where to jump to and continue execution. For any *given* pattern, the compiler knows that that address is the address of the place where that pattern type implements that trait method. But there is no one address we could use for *any* type, so we need to have one copy for each type, each with its own address to jump to. This is referred to as *static dispatch*, since for any given copy of the method, the address we are "dispatching to" is known statically.

NOTE   *You may have noticed that the word "static" is a little overloaded in this context. Static is generally used to refer to anything that is known at compile time, or can be treated as though it were, since it can then be written into static memory, as we discussed in Chapter 1.*

This process of going from a generic type to many non-generic types is called *monomorphization*, and it's part of the reason generic Rust code usually performs just as well as non-generic code. By the time the compiler starts optimizing your code, it's as if no generics were there at all! Each instance is optimized separately and with all of the types known. As a result, the code is just as efficient as if the is_contained_in method of the pattern that is passed in were called directly without any traits present. The compiler has full knowledge of the types involved and can even inline the implementation of is_contained_in if it wishes.

Monomorphization also comes at a cost: all those instantiations of your type need to be compiled separately, which can increase compile time if the compiler cannot optimize them away. Each monomorphized function also results in its own chunk of machine code, which can make your program larger. And because instructions aren't shared between different instantiations of a generic type's methods, the CPU's instruction cache is also less effective as it now needs to hold multiple copies of effectively the same instructions.

NON-GENERIC INNER FUNCTIONS

Often, much of the code in a generic method is not type-dependent. Consider, for example, the implementation of HashMap::insert. The code to compute the hash of the supplied key depends on the key type of the map, but the code to walk the buckets of the map to find the insertion point may not. In cases like this, it would be more efficient to share the generated machine code for the non-generic parts of the method across monomorphizations, and only generate distinct copies where this is actually needed.

One pattern you can use for cases like this is to declare a non-generic helper function inside the generic method that performs the shared operations. This leaves only the type-dependent code for the compiler to copy-paste for you while allowing the helper function to be shared.

Making the function an inner function comes with the added benefit that you do not pollute your module with a single-purpose function. You can instead declare such a helper function outside the method instead; just be careful that you don't make it a method under a generic impl block, as then it will still be monomorphized.

The alternative to static dispatch is *dynamic dispatch*, which enables code to call a trait method on a generic type without knowing what that type is. I said earlier that the reason we needed multiple instances of the method in Listing 2-2 was that otherwise your program wouldn't know what address to jump to in order to call the trait method is_contained_in on the given pattern. Well, with dynamic dispatch, the caller simply tells you. If you replace impl Pattern with &dyn Pattern, you tell the caller that they must give *two* pieces of information for this argument: the address of the pattern *and* the address of the is_contained_in method. In practice, the caller gives us a pointer to a chunk of memory called a virtual method table, or *vtable*, that holds the address of the implementation of *all* the trait's methods for the type in question, one of which is is_contained_in. When the code inside the method wants to call a trait method on the provided pattern, it looks up the address of that pattern's implementation of is_contained_in in the vtable and then calls the function at that address. This allows us to use the same function body regardless of what type the caller wants to use.

NOTE *Every vtable also contains information about the concrete type's layout and alignment since that information is always needed to work with a type. If you want an example of what an explicit vtable looks like, take a look at the std::task::RawWakerVTable type.*

You'll notice that when we opted in to dynamic dispatch using the dyn keyword, we had to place an & in front of it. The reason is that we no longer know at compile time the size of the pattern type that the caller passes in, so we don't know how much space to set aside for it. In other words, dyn Trait is !Sized, where the ! means not. To make it Sized so we can take it as an argument, we place it behind a pointer (which we know the size of). Since we also need to pass along the table of method addresses, this pointer becomes a wide pointer, where the extra word holds the pointer to the vtable. You can use any type that is able to hold a wide pointer for dynamic dispatch, such as &mut, Box, and Arc. Listing 2-3 shows the dynamic dispatch equivalent of Listing 2-2.

```
impl String {
  pub fn contains(&self, p: &dyn Pattern) -> bool {
    p.is_contained_in(&*self)
  }
}
```

Listing 2-3: A generic method using dynamic dispatch

The combination of a type that implements a trait and its vtable is known as a *trait object*. Most traits can be turned into trait objects, but not all. For example, the Clone trait, whose clone method returns Self, cannot be turned into a trait object. If we accept a dyn Clone trait object and then call clone on it, the compiler won't know what type to return. Or, consider the Extend trait from the standard library, which has a method extend that is generic over the type of the provided iterator (so there may be many instances of it). If you were to call a method that took a dyn Extend, there would be no single address for extend to place in the trait object's vtable; there would have to be one entry for every type extend might ever be called with. These are examples of traits that are not *object-safe* and therefore may not be turned into trait objects. To be object-safe, none of a trait's methods can be generic or use the Self type. Furthermore, the trait cannot have any static methods (that is, methods whose first argument does not dereference to Self), since it would be impossible to know which instance of the method to call. It is not clear, for example, what code FromIterator::from_iter(&[0]) should execute.

When reading about trait objects, you may see mentions of the trait bound Self: Sized. Such a bound implies that Self is not being used through a trait object (since it would then be !Sized). You can place that bound on a trait to require that the trait never use dynamic dispatch, or you can place it on a specific method to make that method unavailable when the trait is accessed through a trait object. Methods with a where Self: Sized bound are exempted when checking if a trait is object-safe.

Dynamic dispatch cuts compile times, since it's no longer necessary to compile multiple copies of types and methods, and it can improve the efficiency of your CPU instruction cache. However, it also prevents the compiler from optimizing for the specific types that are used. With dynamic dispatch, all the compiler can do for find in Listing 2-2 is insert a call to the function through the vtable—it can no longer perform any additional optimizations as it does not know what code will sit on the other side of that function call. Furthermore, every method call on a trait object requires a lookup in the vtable, which adds a small amount of overhead over calling the method directly.

When you're given the choice between static and dynamic dispatch, there is rarely a clear-cut right answer. Broadly speaking, though, you'll want to use static dispatch in your libraries and dynamic dispatch in your binaries. In a library, you want to allow your users to decide what kind of dispatch is best for them, since you don't know what their needs are. If you use dynamic dispatch, they're forced to do the same, whereas if you use static dispatch, they can choose whether to use dynamic dispatch or not. In a binary, on the other hand, you're writing the final code, so there are no needs to consider except those of the code you are writing. Dynamic dispatch often allows you to write cleaner code that leaves out generic parameters and will compile more quickly, all at a (usually) marginal performance cost, so it's usually the better choice for binaries.

Generic Traits

Rust traits can be generic in one of two ways: with generic type parameters like trait Foo<T> or with associated types like trait Foo { type Bar; }. The difference between these is not immediately apparent, but luckily the rule of thumb is quite simple: use an associated type if you expect only one implementation of the trait for a given type, and use a generic type parameter otherwise.

The rationale for this is that associated types are often significantly easier to work with, but will not allow multiple implementations. So, more simply put, the advice is really just to use associated types whenever you can.

With a generic trait, users must always specify all the generic parameters and repeat any bounds on those parameters. This can quickly get messy and hard to maintain. If you add a generic parameter to a trait, all users of that trait must also be updated to reflect the change. And since multiple implementations of a trait may exist for a given type, the compiler may have a hard time deciding which instance of the trait you meant to use, leading to awful disambiguating function calls like FromIterator::<u32>::from_iter. But the upside is that you can implement the trait multiple times for the same type—for example, you can implement PartialEq against multiple right-hand side types for your type, or you can implement both FromIterator<T> *and* FromIterator<&T> where T: Clone, precisely because of the flexibility that generic traits provide.

With associated types, on the other hand, the compiler needs to know only the type that implements the trait, and all the associated types follow (since there is only one implementation). This means the bounds can all live in the trait itself and do not need to be repeated on use. In turn, this allows the trait to add further associated types without affecting its users. And because the type dictates all the associated types of the trait, you never have to disambiguate with the unified function calling syntax shown in the previous paragraph. However, you cannot implement Deref against multiple Target types, nor can you implement Iterator with multiple different Item types.

Coherence and the Orphan Rule

Rust has some fairly strict rules about where you can implement traits and what types you can implement them on. These rules exist to preserve the *coherence* property: for any given type and method, there is only ever one correct choice for which implementation of the method to use for that type. To see why this is important, consider what would happen if I could write my own implementation of the `Display` trait for the `bool` type from the standard library. Now, for any code that tries to print a `bool` value and includes my crate, the compiler won't know whether to pick the implementation I wrote or the one from the standard library. Neither choice is correct or better than the other, and the compiler obviously cannot choose randomly. The same issue occurs if the standard library is not involved at all, but we instead have two crates that depend on each other, and they both implement a trait for some shared type. The coherence property ensures that the compiler never ends up in these situations and never has to make these choices: there will always be exactly one obvious choice.

A facile way to uphold coherence would be to ensure only the crate that defines a trait can write implementations for that trait; if no one else can implement the trait, then there can be no conflicting implementations elsewhere. However, this is too restrictive in practice and would essentially make traits useless, as there would be no way to implement traits like `std::fmt::Debug` and `serde::Serialize` for your own types, unless you got your own type included into the defining crate. The opposite extreme, saying that you can implement traits for only your own types, solves that problem but introduces another: a crate that defines a trait now cannot provide implementations of that trait for types in the standard library or in other popular crates! Ideally, we would like to find some set of rules that balances the desire for downstream crates to implement upstream traits for their own types against the desire for upstream crates to be able to add implementations of their own traits without breaking downstream code.

NOTE *Upstream refers to something your code depends on, and downstream refers to something that depends on your code. Often, these terms are used in the direct sense of crate dependencies, but they can also be used to refer to authoritative forks of a codebase—if you do a fork of the Rust compiler, the official Rust compiler is your "upstream."*

In Rust, the rule that establishes that balance is the *orphan rule*. Simply stated, the orphan rule says that you can implement a trait for a type only if the trait *or* the type is local to your crate. So, you can implement `Debug` for your own type, and you can implement `MyNeatTrait` for `bool`, but you cannot implement `Debug` for `bool`. If you try, your code will not compile, and the compiler will tell you that there are conflicting implementations.

This gets you pretty far; it allows you to implement your own traits for third-party types and to implement third-party traits for your own types. However, the orphan rule is not the end of the story. There are a number of additional implications, caveats, and exceptions to it that you should be aware of.

Blanket Implementations

The orphan rule allows you to implement traits over a range of types with code like `impl<T> MyTrait for T where T:` and so on. This is a *blanket implementation*—it is not limited to just one particular type but instead applies to a wide range of types. Only the crate that defines a trait is allowed to write a blanket implementation, and adding a blanket implementation to an existing trait is considered a breaking change. If it were not, a downstream crate that contained `impl MyTrait for Foo` could suddenly stop compiling just because you update the crate that defines `MyTrait` with an error about a conflicting implementation.

Fundamental Types

Some types are so essential that it's necessary to allow anyone to implement traits on them, even if this seemingly violates the orphan rule. These types are marked with the `#[fundamental]` attribute and currently include `&`, `&mut`, and `Box`. For the purposes of the orphan rule, fundamental types may as well not exist—they are effectively erased before the orphan rule is checked in order to allow you to, for example, implement `IntoIterator for &MyType`. With just the orphan rule, this implementation would not be permitted since it implements a foreign trait for a foreign type—`IntoIterator` and `&` both come from the standard library. Adding a blanket implementation over a fundamental type is also considered a breaking change.

Covered Implementations

There are some limited cases where we want to allow implementing a foreign trait for a foreign type, which the orphan rule does not normally allow. The simplest example of this is when you want to write something like `impl From<MyType> for Vec<i32>`. Here, the `From` trait is foreign, as is the `Vec` type, yet there is no danger of violating coherence. This is because a conflicting implementation could be added only through a blanket implementation in the standard library (the standard library cannot otherwise name `MyType`), which is a breaking change anyway.

To allow these kinds of implementations, the orphan rule includes a narrow exemption that permits implementing foreign traits for foreign types under a very specific set of circumstances. Specifically, a given `impl<P1..=Pn> ForeignTrait<T1..=Tn> for T0` is allowed only if at least one `Ti` is a local type and no `T` before the first such `Ti` is one of the generic types `P1..=Pn`. Generic type parameters (`Ps`) *are* allowed to appear in `T0..Ti` as long as they are *covered* by some intermediate type. A `T` is covered if it appears as a type parameter to some other type (like `Vec<T>`), but not if it stands on its own (just `T`) or just appears behind a fundamental type like `&T`. So, all the implementations in Listing 2-4 are valid.

```
impl<T> From<T> for MyType
impl<T> From<T> for MyType<T>
impl<T> From<MyType> for Vec<T>
impl<T> ForeignTrait<MyType, T> for Vec<T>
```

Listing 2-4: Valid implementations of foreign traits for foreign types

However, the implementations in Listing 2-5 are invalid.

```
impl<T> ForeignTrait for T
impl<T> From<T> for T
impl<T> From<Vec<T>> for T
impl<T> From<MyType<T>> for T
impl<T> From<T> for Vec<T>
impl<T> ForeignTrait<T, MyType> for Vec<T>
```

Listing 2-5: Invalid implementations of foreign traits for foreign types

This relaxation of the orphan rule complicates the rules for what con-stitutes a breaking change when you add a new implementation for an exist-ing trait. In particular, adding a new implementation to an existing trait is non-breaking only if it contains at least one *new* local type, and that new local type satisfies the rules for the exemption described earlier. Adding any other new implementation is a breaking change.

NOTE   *Note that impl<T> ForeignTrait<LocalType, T> for ForeignType is valid, but impl<T> ForeignTrait<T, LocalType> for ForeignType is not! This may seem arbi-trary, but without this rule, you could write impl<T> ForeignTrait<T, LocalType> for ForeignType, and another crate could write impl<T> ForeignTrait<TheirType, T> for ForeignType, and a conflict would arise only when the two crates were brought together. Instead of disallowing this pattern altogether, the orphan rule requires that your local type come before the type parameter, which breaks the tie and ensures that if both crates uphold coherence in isolation, they also uphold it when combined.*

Trait Bounds

The standard library is flush with trait bounds, whether it's that the keys in a HashMap must implement Hash + Eq or that the function given to thread::spawn must be FnOnce + Send + 'static. When you write generic code yourself, it will almost certainly include trait bounds, as otherwise your code cannot do much with the type it is generic over. As you write more elaborate generic implementations, you'll find that you also need more fidelity from your trait bounds, so let's look at some of the ways to achieve that.

First and foremost, trait bounds do not have to be of the form T: Trait where T is some type your implementation or type is generic over. The bounds can be arbitrary type restrictions and do not even need to include generic parameters, types of arguments, or local types. You can write a trait bound like where String: Clone, even though String: Clone is always true and contains no local types. You can also write where io::Error: From<MyError<T>>; your generic type parameters do not need to appear only on the left-hand side. This not only allows you to express more intricate bounds but also can save you from needlessly repeating bounds. For example, if your method wants to construct a HashMap<K, V, S> whose keys are some generic type T and whose value is a usize, instead of writing the bounds out like where T: Hash + Eq, S: BuildHasher + Default, you could write where HashMap<T, usize, S>: FromIterator. This saves you from looking up the exact bounds requirements for the methods you end up using and more clearly communicates the "true" requirement of your code. As you can see, it can also significantly reduce the complexity of your bounds if the bounds on the underlying trait methods you want to call are complex.

DERIVE TRAIT

While `#[derive(Trait)]` is extremely convenient, in the context of trait bounds, you should be aware of one subtlety around how it is often implemented. Many `#[derive(Trait)]` expansions desugar into `impl Trait for Foo<T>` where `T: Trait`. This is often what you want, but not always. For example, consider what happens if we try to derive `Clone` this way for `Foo<T>` and `Foo` contains an `Arc<T>`. `Arc` implements `Clone` *regardless* of whether `T` implements `Clone`, but due to the derived bounds, `Foo` will implement `Clone` only if `T` does! This isn't usually too big of an issue, but it does add a bound where one isn't needed. If we rename the type to `Shared`, the problem may become a little clearer. Imagine how confused a user that has a `Shared<NotClone>` will be when the compiler tells them that they cannot clone it! At the time of writing, this is how `#[derive(Clone)]` as provided by the standard library works, though this may change in the future.

Sometimes, you want bounds on associated types of types you're generic over. As an example, consider the iterator method `flatten`, which takes an iterator that produces items that in turn implement `Iterator` and produces an iterator of the items of those inner iterators. The type it produces, `Flatten`, is generic over `I`, which is the type of the outer iterator. `Flatten` implements `Iterator` if `I` implements `Iterator` *and* the items yielded by `I` themselves implement `IntoIterator`. To enable you to write bounds like this, Rust lets you refer to associated types of a type using the syntax `Type::AssocType`. For example, we can refer to `I`'s `Item` type using `I::Item`. If a type has multiple associated types by the same name, such as if the trait that provides the associated type is itself generic (and therefore there are many implementations), you can disambiguate with the syntax `<Type as Trait>::AssocType`. Using this, you can write bounds not only for the outer iterator type but also for the item type of that outer iterator.

In code that uses generics extensively, you may find that you need to write a bound that talks about references to a type. This is normally fine, as you'll tend to also have a generic lifetime parameter that you can use as the lifetime for these references. In some cases, however, you want the bound to say "this reference implements this trait for any lifetime." This type of bound is known as a *higher-ranked trait bound*, and it's particularly useful in association with the `Fn` traits. For example, say you want to be generic over a function that takes a reference to a `T` and returns a reference to *inside* that `T`. If you write `F: Fn(&T) -> &U`, you need to provide a lifetime for those references, but you really want to say "any lifetime as long as the output is the same as the input." Using a higher-ranked lifetime, you can write F: for<'a> Fn(&'a T) -> &'a U to say that for *any* lifetime 'a, the bound must hold. The Rust compiler is smart enough that it automatically adds the for when you write Fn bounds with references like this, which covers the majority of use cases for this feature. The explicit form is needed so exceedingly rarely that, at the time of writing, the standard library uses it in just three places—but it does happen and so is worth knowing about.

To bring all of this together, consider the code in Listing 2-6, which can be used to implement Debug for any type that can be iterated over and whose elements are Debug.

```
impl Debug for AnyIterable
  where for<'a> &'a Self: IntoIterator,
        for<'a> <&'a Self as IntoIterator>::Item: Debug {
    fn fmt(&self, f: &mut Formatter) -> Result<(), Error> {
        f.debug_list().entries(self).finish()
}}
```

Listing 2-6: An excessively generic implementation of Debug for any iterable collection

You could copy-paste this implementation for pretty much any collection type and it would "just work." Of course, you may want a smarter debug implementation, but this illustrates the power of trait bounds quite well.

Marker Traits

Usually, we use traits to denote functionality that multiple types can support; a Hash type can be hashed by calling hash, a Clone type can be cloned by calling clone, and a Debug type can be formatted for debugging by calling fmt. But not all traits are functional in this way. Some traits, called *marker traits*, instead indicate a property of the implementing type. Marker traits have no methods or associated types and serve just to tell you that a particular type can or cannot be used in a certain way. For example, if a type implements the Send marker trait, it is safe to send across thread boundaries. If it does not implement this marker trait, it isn't safe to send. There are no methods associated with this behavior; it's just a fact about the type. The standard library has a number of these in the std::marker module, including Send, Sync, Copy, Sized, and Unpin. Most of these (all except Copy) are also *auto-traits*; the compiler automatically implements them for types unless the type contains something that does not implement the marker trait.

Marker traits serve an important purpose in Rust: they allow you to write bounds that capture semantic requirements not directly expressed in the code. There is no call to send in code that requires that a type is Send. Instead, the code *assumes* that the given type is fine to use in a separate thread, and without marker traits the compiler would have no way of checking that assumption. It would be up to the programmer to remember the assumption and read the code very carefully, which we all know is not something we'd like to rely on. That path is riddled with data races, segfaults, and other runtime issues.

Similar to marker traits are *marker types*. These are unit types (like struct MyMarker;) that hold no data and have no methods. Marker types are useful for, well, marking a type as being in a particular state. They come in handy when you want to make it impossible for a user to misuse an API. For example, consider a type like SshConnection, which may or may not have been authenticated yet. You could add a generic type argument to SshConnection and then create two marker types: Unauthenticated and Authenticated. When the user first connects, they get SshConnection<Unauthe nticated>. In its impl block, you provide only a single method: connect. The connect method returns a SshConnection<Authenticated>, and it's only in that impl block that you provide the remaining methods for running commands and such. We will look at this pattern further in Chapter 3.

Existential Types

In Rust you very rarely have to specify the types of variables you declare in the body of a function or the types of generic arguments to methods that you call. This is because of *type inference*, where the compiler decides what type to use based on what type the code the type appears in evaluates to. The compiler will usually infer types only for variables and for the arguments (and return types) of closures; top-level definitions like functions, types, traits, and trait implementation blocks all require that you explicitly name all types. There are a couple of reasons for this, but the primary one is that type inference is much easier when you have at least some known points to start the inference from. However, it's not always easy, or even possible, to fully name a type! For example, if you return a closure from a function, or an async block from a trait method, its type does not have a name that you can type into your code.

To handle situations like this, Rust supports *existential types*. Chances are, you have already seen existential types in action. All functions marked as async fn or with a return type of impl Trait have an existential return type: the signature does not give the true type of the return value, just a hint that the function returns *some* type that implements some set of traits that the caller can rely on. And crucially, the caller can only rely on the return type implementing those traits, and nothing else.

NOTE *Technically, it isn't strictly true that the caller relies on the return type and nothing else. The compiler will also propagate auto-traits like Send and Sync through impl Trait in return position. We'll look at this more in the next chapter.*

This behavior is what gives existential types their name: we are asserting that there exists some concrete type that matches the signature, and we leave it up to the compiler to find what that type is. The compiler will usually then go figure that out by applying type inference on the body of the function.

Not all instances of `impl Trait` use existential types. If you use `impl Trait` in argument position for a function, it's really just shorthand for an unnamed generic parameter to that function. For example, `fn foo(s: impl ToString)` is mostly just syntax sugar for `fn foo<S: ToString>(s: S)`.

Existential types come in handy particularly when you implement traits that have associated types. For example, imagine you're implementing the `IntoIterator` trait. It has an associated type `IntoIter` that holds the type of the iterator that the type in question can be turned into. With existential types, you do not need to define a separate iterator type to use for `IntoIter`. Instead, you can give the associated type as `impl Iterator<Item = Self::Item>` and just write an expression inside the `fn into_iter(self)` that evaluates to an `Iterator`, such as by using maps and filters over some existing iterator type.

Existential types also provide a feature beyond mere convenience: they allow you to perform zero-cost type erasure. Instead of exporting helper types just because they appear in a public signature somewhere—iterators and futures are common examples of this—you can use existential types to hide the underlying concrete type. Users of your interface are shown only the traits that the relevant type implements, while the concrete type is left as an implementation detail. Not only does this simplify the interface, but it also enables you to change that implementation as you wish without breaking downstream code in the future.

Summary

This chapter has provided a thorough review of the Rust type system. We've looked both at how the compiler manifests types in memory and how it reasons about the types themselves. This is important background material for writing unsafe code, complex application interfaces, and asynchronous code in later chapters. You'll also find that much of the type reasoning from this chapter plays into how you design Rust code interfaces, which we'll cover in the next chapter.

3

DESIGNING INTERFACES

Every project, no matter how large or small, has an API. In fact, it usually has several. Some of these are user-facing, like an HTTP endpoint or a command line interface, and some are developer-facing, like a library's public interface. On top of these, Rust crates also have a number of internal interfaces: every type, trait, and module boundary has its own miniature API that the rest of your code interfaces with. As your codebase grows in size and complexity, you'll find it worthwhile to invest some thought and care into how you design even the internal APIs to make the experience of using and maintaining the code over time as pleasant as possible.

In this chapter we'll look at some of the most important considerations for writing idiomatic interfaces in Rust, whether the users of those interfaces are your own code or other developers using your library. These essentially boil down to four principles: your interfaces should be *unsurprising*, *flexible*, *obvious*, and *constrained*. I'll discuss each of these principles in turn, to provide some guidance for writing reliable and usable interfaces.

I highly recommend taking a look at the Rust API Guidelines (*https://rust-lang.github.io/api-guidelines/*) after you've read this chapter. There's an excellent checklist you can follow, with a detailed run-through of each recommendation. Many of the recommendations in this chapter are also checked by the `cargo clippy` tool, which you should start running on your code if you aren't already. I also encourage you to read through Rust RFC 1105 (*https://rust-lang.github.io/rfcs/1105-api-evolution.html*) and the chapter of *The Cargo Book* on SemVer compatibility (*https://doc.rust-lang.org/cargo/reference/semver.html*), which cover what is and is not a breaking change in Rust.

Unsurprising

The Principle of Least Surprise, otherwise known as the Law of Least Astonishment, comes up a lot in software engineering, and it holds true for Rust interfaces as well. Where possible, your interfaces should be intuitive enough that if the user has to guess, they usually guess correctly. Of course, not everything about your application is going to be immediately intuitive in this way, but anything that *can* be unsurprising should be. The core idea here is to stick close to things the user is likely to already know so that they don't have to relearn concepts in a different way than they're used to. That way you can save their brain power for figuring out the things that are actually specific to your interface.

There are a variety of ways you can make your interfaces predictable. Here, we'll look at how you can use naming, common traits, and ergonomic trait tricks to help the user out.

Naming Practices

A user of your interface will encounter it first through its names; they will immediately start to infer things from the names of types, methods, variables, fields, and libraries they come across. If your interface reuses names for things—say, methods and types—from other (perhaps common) interfaces, the user will know they can make certain assumptions about your methods and types. A method called `iter` probably takes `&self`, and probably gives you an iterator. A method called `into_inner` probably takes `self` and likely returns some kind of wrapped type. A type called `SomethingError` probably implements `std::error::Error` and appears in various `Results`. By reusing common names for the same purpose, you make it easier for the user to guess what things do and allow them to more easily understand the things that are different about your interface.

A corollary to this is that things that share a name *should* in fact work the same way. Otherwise—for example, if your `iter` method takes `self`, or if your `SomethingError` type does not implement `Error`—the user will likely write incorrect code based on how they expect the interface to work. They will be surprised and frustrated and will have to spend time digging into how your interface differs from their expectations. When we can save the user this kind of friction, we should.

Common Traits for Types

Users in Rust will also make the major assumption that everything in the interface "just works." They expect to be able to print any type with {:?} and send anything and everything to another thread, and they expect that every type is Clone. Where possible, we should again avoid surprising the user and eagerly implement most of the standard traits even if we do not need them immediately.

Because of the coherence rules discussed in Chapter 2, the compiler will not allow users to implement these traits when they need them. Users aren't allowed to implement a foreign trait (like Clone) for a foreign type like one from your interface. They would instead need to wrap your interface type in their own type, and even then it may be quite difficult to write a reasonable implementation without access to the type's internals.

First among these standard traits is the Debug trait. Nearly every type can, and should, implement Debug, even if it only prints the type's name. Using #[derive(Debug)] is often the best way to implement the Debug trait in your interface, but keep in mind that all derived traits automatically add the same bound for any generic parameters. You could also simply write your own implementation by leveraging the various debug_ helpers on fmt::Formatter.

Tied in close second are the Rust auto-traits Send and Sync (and, to a lesser extent, Unpin). If a type does not implement one of these traits, it should be for a very good reason. A type that is not Send can't be placed in a Mutex and can't be used even transitively in an application that contains a thread pool. A type that is not Sync can't be shared through an Arc or placed in a static variable. Users have come to expect that types *just work* in these contexts, especially in the asynchronous world where nearly everything runs on a thread pool, and will become frustrated if you don't ensure that your types implement these traits. If your types cannot implement them, make sure that fact, and the reason why, is well documented!

The next set of nearly universal traits you should implement is Clone and Default. These traits can be derived or implemented easily and make sense to implement for most types. If your type cannot implement these traits, make sure to call it out in your documentation, as users will usually expect to be able to easily create more (and new) instances of types as they see fit. If they cannot, they will be surprised.

One step further down in the hierarchy of expected traits is the comparison traits: PartialEq, PartialOrd, Hash, Eq, and Ord. The PartialEq trait is particularly desirable, because users will at some point inevitably have two instances of your type that they wish to compare with == or assert_eq!. Even if your type would compare equal for only the same instance of the type, it's worth implementing PartialEq to enable your users to use assert_eq!.

PartialOrd and Hash are more specialized, and may not apply quite as broadly, but where possible you will want to implement them too. This is especially true for types a user might use as the key in a map, or a type they may deduplicate using any of the std::collection set types, since they tend to require these bounds. Eq and Ord come with additional semantic requirements on the implementing type's comparison operations beyond those of PartialEq and PartialOrd. These are well documented in the documentation for those traits, and you should implement them *only* if you're sure those semantics actually apply to your type.

Finally, for most types, it makes sense to implement the serde crate's Serialize and Deserialize traits. These can be easily derived, and the serde _derive crate even comes with mechanisms for overwriting the serialization for just one field or enum variant. Since serde is a third-party crate, you may not wish to add a required dependency on it. Most libraries therefore choose to provide a serde feature that adds support for serde only when the user opts into it.

You might be wondering why I haven't included the derivable trait Copy in this section. There are two things that set Copy apart from the other traits mentioned. The first is that users do not generally expect types to be Copy; quite to the contrary, they tend to expect that if they want two copies of something, they have to call clone. Copy changes the semantics of moving a value of the given type, which might surprise the user. This ties in to the second observation: it is very easy for a type to *stop* being Copy, because Copy types are highly restricted. A type that starts out simple can easily end up having to hold a String, or some other non-Copy type. Should that happen, and you have to remove the Copy implementation, that's a backward incompatible change. In contrast, you rarely have to remove a Clone implementation, so that's a less onerous commitment.

Ergonomic Trait Implementations

Rust does not automatically implement traits for references to types that implement traits. To phrase this a different way, you cannot generally call fn foo<T: Trait>(t: T) with a &Bar, even if Bar: Trait. This is because Trait may contain methods that take &mut self or self, which obviously cannot be called on &Bar. Nonetheless, this behavior might be very surprising to a user who sees that Trait has only &self methods!

For this reason, when you define a new trait, you'll usually want to provide blanket implementations as appropriate for that trait for &T where T: Trait, &mut T where T: Trait, and Box<T> where T: Trait. You may be able to implement only some of these depending on what receivers the methods of Trait have. Many of the traits in the standard library have similar implementations, precisely because that leads to fewer surprises for the user.

Iterators are another case where you'll often want to specifically add trait implementations on references to a type. For any type that can be iterated over, consider implementing IntoIterator for both &MyType and &mut MyType where applicable. This makes for loops work with borrowed instances of your type as well out of the box, just like users would expect.

Wrapper Types

Rust does not have object inheritance in the classical sense. However, the Deref trait and its cousin AsRef both provide something a little like inheritance. These traits allow you to have a value of type T and call methods on some type U by calling them directly on the T-typed value if T: Deref<Target = U>. This feels like magic to the user, and is generally great.

If you provide a relatively transparent wrapper type (like Arc), there's a good chance you'll want to implement Deref so that users can call methods on the inner type by just using the . operator. If accessing the inner type does not require any complex or potentially slow logic, you should also consider implementing AsRef, which allows users to easily use a &WrapperType as an &InnerType. For most wrapper types, you will also want to implement From<InnerType> and Into<InnerType> where possible so that your users can easily add or remove your wrapping.

You may also have come across the Borrow trait, which feels very similar to Deref and AsRef but is really a bit of a different beast. Specifically, Borrow is tailored for a much narrower use case: allowing the caller to supply any one of multiple essentially identical variants of the same type. It could, perhaps, have been called Equivalent instead. For example, for a HashSet<String>, Borrow allows the caller to supply either a &str or a &String. While the same could have been achieved with AsRef, that would not be safe without Borrow's additional requirement that the target type implements Hash, Eq, and Ord exactly the same as the implementing type. Borrow also has a blanket implementation of Borrow<T> for T, &T, and &mut T, which makes it convenient to use in trait bounds to accept either owned *or* referenced values of a given type. In general, Borrow is intended only for when your type is essentially equivalent to another type, whereas Deref and AsRef are intended to be implemented more widely for anything your type can "act as."

DEREF AND INHERENT METHODS

The magic around the dot operator and Deref can get confusing and surprising when there are methods on T that take self. For example, given a value t: T, it is not clear whether t.frobnicate() frobnicates the T or the underlying U!

For this reason, types that allow you to transparently call methods on some inner type that isn't known in advance should avoid inherent methods. It's fine for Vec to have a push method even though it dereferences to a slice, since you know that slices won't get a push method any time soon. But if your type dereferences to a user-controlled type, any inherent method you add may also exist on that user-controlled type, and thus cause issues. In these cases, favor static methods of the form fn frobnicate(t: T). That way, t.frobnicate() always calls U::frobnicate, and T::frobnicate(t) can be used to frobnicate the T itself.

Flexible

Every piece of code you write includes, implicitly or explicitly, a contract. The contract consists of a set of requirements and a set of promises. The requirements are restrictions on how the code can be used, while the promises are guarantees about how the code can be used. When designing a new interface, you want to think carefully about this contract. A good rule of thumb is to avoid imposing unnecessary restrictions and to only make promises you can keep. Adding restrictions or removing promises usually requires a major semantic version change and is likely to break code elsewhere. Relaxing restrictions or giving additional promises, on the other hand, is usually backward compatible.

In Rust, restrictions usually come in the form of trait bounds and argument types, and promises come in the form of trait implementations and return types. For example, compare the three function signatures in Listing 3-1.

```
fn frobnicate1(s: String) -> String
fn frobnicate2(s: &str) -> Cow<'_, str>
fn frobnicate3(s: impl AsRef<str>) -> impl AsRef<str>
```

Listing 3-1: Similar function signatures with different contracts

These three function signatures all take a string and return a string, but they do so under very different contracts.

The first function requires the caller to own the string in the form of the String type, and it promises that it will return an owned String. Since the contract requires the caller to allocate and requires us to return an owned String, we cannot later make this function allocation-free in a backward compatible way.

The second function relaxes the contract: the caller can provide any reference to a string, so the user no longer needs to allocate or give up ownership of a String. It also promises to give back a std::borrow::Cow, meaning it can return either a string reference or an owned String, depending on whether it needs to own the string. The promise here is that the function will always return a Cow, which means that we cannot, say, change it to use some other optimized string representation later. The caller must also specifically provide a &str, so if they have, say, a pre-existing String of their own, they must dereference it to a &str to call our function.

The third function lifts these restrictions. It requires only that the user pass in a type that can produce a reference to a string, and it promises only that the return value can produce a reference to a string.

None of these function signatures is *better* than the others. If you need ownership of a string in the function, you can use the first argument type to avoid an extra string copy. If you want to allow the caller to take advantage of the case where an owned string was allocated and returned, the second function with a return type of Cow may be a good choice. Instead, what I want you to take away from this is that you should think carefully about what contract your interface binds you to, because changing it after the fact can be disruptive.

In the remainder of this section I give examples of interface design decisions that often come up, and their implications for your interface contract.

Generic Arguments

One obvious requirement your interface must place on users is what types they must provide to your code. If your function explicitly takes a `Foo`, the user must own and give you a `Foo`. There is no way around it. In most cases it pays off to use generics rather than concrete types, to allow the caller to pass any type that conforms to what your function actually needs, rather than only a particular type. Changing `&str` in Listing 3-1 to `impl AsRef<str>` is an example of this kind of relaxing. One way to go about relaxing requirements this way is to start with the argument fully generic with no bounds, and then just follow the compiler errors to discover what bounds you need to add.

However, if taken to the extreme, this approach would make every argument to every function its own generic type, which would be both hard to read and hard to understand. There are no hard-and-fast rules for exactly when you should or should not make a given parameter generic, so use your best judgment. A good rule of thumb is to make an argument generic if you can think of other types a user might reasonably and frequently want to use instead of the concrete type you started with.

You may remember from Chapter 2 that generic code is duplicated for every combination of types ever used with the generic code through monomorphization. With that in mind, the idea of making lots of arguments generic might make you worried about overly enlarging your binaries. In Chapter 2 we also discussed how you can use dynamic dispatch to mitigate this at a (usually) negligible performance cost, and that applies here too. For arguments that you take by reference anyway (recall that `dyn Trait` is not `Sized`, and that you need a wide pointer to use them), you can easily replace your generic argument with one that uses dynamic dispatch. For instance, instead of `impl AsRef<str>`, you could take `&dyn AsRef<str>`.

Before you go running to do that, though, there are a few things you should consider. First, you are making this choice on behalf of your users, who cannot opt out of dynamic dispatch. If you know that the code you're applying dynamic dispatch to will never be performance-sensitive, that may be fine. But if a user comes along who wants to use your library in their high-performance application, dynamic dispatch in a function that is called in a hot loop may be a deal breaker. Second, at the time of writing, using dynamic dispatch will work only when you have a simple trait bound like `T: AsRef<str>` or `impl AsRef<str>`. For more complex bounds, Rust does not know how to construct a dynamic dispatch vtable, so you cannot take, say, `&dyn Hash + Eq`. And finally, remember that with generics, the caller can always choose dynamic dispatch themselves by passing in a trait object. The reverse is not true: if you take a trait object, that is what the caller must provide.

It may be tempting to start your interfaces off with concrete types and then turn them generic over time. This can work, but keep in mind that such changes are not necessarily backward compatible. To see why, imagine that you change a function from `fn foo(v: &Vec<usize>)` to `fn foo(v: impl AsRef<[usize]>)`. While every `&Vec<usize>` implements `AsRef<[usize]>`, type inference can still cause issues for users. Consider what happens if the caller invokes foo with foo(&iter.collect()). In the original version, the compiler could determine that it should collect into a Vec, but now it just knows that it needs to collect into some type that implements AsRef<[usize]>. And there could be multiple such types, so with this change, the caller's code will no longer compile!

Object Safety

When you define a new trait, whether or not that trait is object-safe (see the end of "Compilation and Dispatch" in Chapter 2) is an unwritten part of the trait's contract. If the trait is object-safe, users can treat different types that implement your trait as a single common type using dyn Trait. If it isn't, the compiler will disallow dyn Trait for that trait. You should prefer your traits to be object-safe even if that comes at a slight cost to the ergonomics of using them (such as taking impl AsRef<str> over &str), since object safety enables new ways to use your traits. If your trait must have a generic method, consider whether its generic parameters can be on the trait itself or if its generic arguments can also use dynamic dispatch to preserve the object safety of the trait. Alternatively, you can add a where Self: Sized trait bound to that method, which makes it possible to call the method only with a concrete instance of the trait (and not through dyn Trait). You can see examples of this pattern in the Iterator and Read traits, which are object-safe but provide some additional convenience methods on concrete instances.

There is no single answer to the question of how many sacrifices you should be willing to make to preserve object safety. My recommendation is that you consider how your trait will be used, and whether it makes sense for users to want to use it as a trait object. If you think it's likely that users will want to use many different instances of your trait together, you should work harder to provide object safety than if you don't think that use case makes much sense. For example, dynamic dispatch would not be useful for the FromIterator trait because its one method does not take self, so you wouldn't be able to construct a trait object in the first place. Similarly, std::io::Seek is fairly useless as a trait object on its own, because the only thing you would be able to do with such a trait object is seek, without being able to read or write.

DROP TRAIT OBJECTS

You might think that the Drop trait is also useless as a trait object, since all you can do with Drop as a trait object is to drop it. But it turns out there are some libraries that specifically just want to be able to drop arbitrary types. For example, a library that offers deferred dropping of values, such as for concurrent garbage collection or just deferred cleanup, cares only that the values can

---

> be dropped, and nothing else. Interestingly enough, the story of Drop doesn't end there; since Rust needs to be able to drop trait objects too, *every* vtable contains the drop method. Effectively, every dyn Trait is also a dyn Drop.

Remember that object safety is a part of your public interface! If you modify a trait in an otherwise backward compatible way, such as by adding a method with a default implementation, but it makes the trait not object-safe, you need to bump your major semantic version number.

Borrowed vs. Owned

For nearly every function, trait, and type you define in Rust, you must decide whether it should own, or just hold a reference to, its data. Whatever decision you make will have far-reaching implications for the ergonomics and performance of your interface. Luckily, these decisions very often make themselves.

If the code you write needs ownership of the data, such as to call methods that take self or to move the data to another thread, it must store the owned data. When your code must own data, it should generally also make the caller provide owned data, rather than taking values by reference and cloning them. This leaves the caller in control of allocation, and it is upfront about the cost of using the interface in question.

On the other hand, if your code doesn't need to own the data, it should operate on references instead. One common exception to this rule is with small types like i32, bool, or f64, which are just as cheap to store and copy directly as to store through references. Be wary of assuming this holds true for all Copy types, though; [u8; 8192] is Copy, but it would be expensive to store and copy it all over the place.

Of course, in the real world, things are often less clear-cut. Sometimes, you don't know in advance whether your code will need to own the data or not. For example, String::from_utf8_lossy needs to take ownership of the byte sequence that is passed to it only if it contains invalid UTF-8 sequences. In this case, the Cow type is your friend: it lets you operate on references if the data allows, and it lets you produce an owned value if necessary.

Other times, reference lifetimes complicate the interface so much that it becomes a pain to use. If your users are struggling to get code to compile on top of your interface, that's a sign that you may want to (even unnecessarily) take ownership of certain pieces of data. If you do this, start with data that is cheap to clone or is not part of anything performance-sensitive before you decide to heap-allocate what might be a huge chunk of bytes.

Fallible and Blocking Destructors

Types centered on I/O often need to perform cleanup when they're dropped. This may include flushing writes to disk, closing files, or gracefully terminating connections to remote hosts. The natural place to perform this cleanup is in the type's `Drop` implementation. Unfortunately, once a value is dropped, we no longer have a way to communicate errors to the user except by panicking. A similar problem arises in asynchronous code, where we wish to finish up when there is work pending. By the time `drop` is called, the executor may be shutting down, and we have no way to do more work. We could try to start another executor, but that comes with its own host of problems, such as blocking in asynchronous code, as we will see in Chapter 8.

There is no perfect solution to these problems, and no matter what we do, some applications will inevitably fall back to our `Drop` implementation. For that reason, we need to provide best-effort cleanup through `Drop`. If cleanup errors, at least we tried—we swallow the error and move on. If an executor is still available, we might spawn a future to do cleanup, but if it never gets to run, we did what we could.

However, we ought to provide a better alternative for users who wish to leave no loose threads. We can do this by providing an explicit destructor. This usually takes the form of a method that takes ownership of `self` and exposes any errors (using `-> Result<_, _>`) or asynchrony (using `async fn`) that are inherent to the destruction. A careful user can then use that method to gracefully tear down any associated resources.

NOTE *Make sure you highlight the explicit destructor in your documentation!*

As always, there's a trade-off. The moment you add an explicit destructor, you will run into two issues. First, since your type implements `Drop`, you can no longer move out of any of that type's fields in the destructor. This is because `Drop::drop` will still be called after your explicit destructor runs, and it takes `&mut self`, which requires that no part of `self` has been moved. Second, `drop` takes `&mut self`, not `self`, so your `Drop` implementation cannot simply call your explicit destructor and ignore its result (because it doesn't own `self`). There are a couple of ways around these problems, none of which are perfect.

The first is to make your top-level type a newtype wrapper around an `Option`, which in turn holds some inner type that holds all of the type's fields. You can then use `Option::take` in both destructors, and call the inner type's explicit destructor only if the inner type has not already been taken. Since the inner type does not implement `Drop`, you can take ownership of all the fields there. The downside of this approach is that all the methods you wish to provide on the top-level type must now include code to get through the `Option` (which you know is always `Some` since `drop` has not yet been called) to the fields on the inner type.

The second workaround is to make each of your fields *takeable.* You can "take" an Option by replacing it with None (which is what Option::take does), but you can do this with many other types as well. For example, you can take a Vec or HashMap by simply replacing them with their cheap-to-construct default values—std::mem::take is your friend here. This approach works great if your types have sane "empty" values but gets tedious if you must wrap nearly every field in an Option and then modify every access of those fields with a matching unwrap.

The third option is to hold the data inside the ManuallyDrop type, which dereferences to the inner type, so there's no need for unwraps. You can also use ManuallyDrop::take in drop to take ownership at destruction time. The primary downside of this approach is that ManuallyDrop::take is unsafe. There are no safety mechanisms in place to ensure that you don't try to use the value inside the ManuallyDrop after you've called take or that you don't call take multiple times. If you do, your program will silently exhibit undefined behavior, and bad things will happen.

Ultimately, you should choose whichever of these approaches fits your application best. I would err on the side of going with the second option, and switching to the others only if you find yourself in a sea of Options. The ManuallyDrop solution is excellent if the code is simple enough that you can easily check the safety of your code, and you are confident in your ability to do so.

Obvious

While some users may be familiar with aspects of the implementation that underpins your interface, they are unlikely to understand all of its rules and limitations. They won't know that it's never okay to call foo after calling bar, or that it's only safe to call the unsafe method baz when the moon is at a 47-degree angle and no one has sneezed in the past 18 seconds. Only if the interface makes it clear that something strange is going on will they reach for the documentation or carefully read type signatures. It's therefore critical for you to make it as easy as possible for users to understand your interface and as hard as possible for them to use it incorrectly. The two primary techniques at your disposal for this are your documentation and the type system, so let's look at each of those in turn.

NOTE *You can also take advantage of naming to suggest to the user when there's more to an interface than meets the eye. If a user sees a method named dangerous, chances are they will read its documentation.*

Documentation

The first step to making your interfaces transparent is to write good documentation. I could write an entire book dedicated to how to write documentation, but let's focus on Rust-specific advice here.

First, clearly document any cases where your code may do something unexpected, or where it relies on the user doing something beyond what's dictated by the type signature. Panics are a good example of both of these circumstances: if your code can panic, document that fact, along with the circumstances it might panic under. Similarly, if your code might return an error, document the cases in which it does. For unsafe functions, document what the caller must guarantee in order for the call to be safe.

Second, include end-to-end usage examples for your code on a crate and module level. These are more important than examples for specific types or methods, since they give the user a feel for how everything fits together. With a decent high-level understanding of the interface's structure, the developer may soon realize what particular methods and types do and where they should be used. End-to-end examples also give the user a starting point for customizing their usage, and they can, and often will, copy-paste the example and then modify it to suit their needs. This kind of "learning by doing" tends to work better than having them try to piece something together from the components.

NOTE *Very method-specific examples that show that, yes, the* len *method indeed returns the length are unlikely to tell the user anything new about your code.*

Third, organize your documentation. Having all your types, traits, and functions in a single top-level module makes it difficult for the user to get a sense of where to start. Take advantage of modules to group together semantically related items. Then, use intra-documentation links to interlink items. If the documentation on type A talks about trait B, then it should link to that trait right there. If you make it easy for the user to explore your interface, they are less likely to miss important connections or dependencies. Also consider marking parts of your interface that are not intended to be public but are needed for legacy reasons with #[doc(hidden)], so that they do not clutter up your documentation.

And finally, enrich your documentation wherever possible. Link to external resources that explain concepts, data structures, algorithms, or other aspects of your interface that may have good explanations elsewhere. RFCs, blog posts, and whitepapers are great for this, if any are relevant. Use #[doc(cfg(..))] to highlight items that are available only under certain configurations so the user quickly realizes why some method that's listed in the documentation isn't available. Use #[doc(alias = "...")] to make types and methods discoverable under other names that users may search for them by. In the top-level documentation, point the user to commonly used modules, features, types, traits, and methods.

Type System Guidance

The type system is an excellent tool to ensure that your interfaces are obvious, self-documenting, and misuse-resistant. You have several techniques at your disposal that can make your interfaces very hard to misuse, and thus, make it more likely that they will be used correctly.

The first of these is *semantic typing*, in which you add types to represent the *meaning* of a value, not just its primitive type. The classic example here is for Booleans: if your function takes three bool arguments, chances are some user will mess up the order of the values and realize it only after something has gone terribly wrong. If, on the other hand, it takes three arguments of distinct two-variant enum types, the user cannot get the order wrong without the compiler yelling at them: if they attempt to pass DryRun::Yes to the overwrite argument, that will simply not work, nor will passing Overwrite::No as the dry_run argument. You can apply semantic typing beyond Booleans as well. For example, a newtype around a numeric type may provide a unit for the contained value, or it could constrain raw pointer arguments to only those that have been returned by another method.

A closely related technique is to use zero-sized types to indicate that a particular fact is true about an instance of a type. Consider, for instance, a type called Rocket that represents the state of a real rocket. Some operations (methods) on Rocket should be available no matter what state the rocket is in, but some make sense only in particular situations. It is, for example, impossible to launch a rocket if it has already been launched. Similarly, it should probably not be possible to separate the fuel tank if the rocket has not yet launched. We could model these as enum variants, but then all the methods would be available at every stage, and we'd need to introduce possible panics.

Instead, as shown in Listing 3-2, we can introduce a generic parameter on Rocket, Stage, and use it to restrict what methods are available when.

```
❶ struct Grounded;
  struct Launched;
  // and so on
  struct Rocket<Stage = Grounded> {
  ❷ stage: std::marker::PhantomData<Stage>,
  }

❸ impl Default for Rocket<Grounded> {}
  impl Rocket<Grounded> {
      pub fn launch(self) -> Rocket<Launched> { }
  }
❹ impl Rocket<Launched> {
      pub fn accelerate(&mut self) { }
      pub fn decelerate(&mut self) { }
  }

❺ impl<Stage> Rocket<Stage> {
      pub fn color(&self) -> Color { }
      pub fn weight(&self) -> Kilograms { }
  }
```

Listing 3-2: Using marker types to restrict implementations

We introduce unit types to represent each stage of the rocket ❶. We don't actually need to store the stage—only the meta-information it provides— so we store it behind a PhantomData ❷ to guarantee that it is eliminated at compile time. Then, we write implementation blocks for Rocket only when it holds a particular type parameter. You can construct a rocket only on the

ground (for now), and you can launch it only from the ground ❸.
Only when the rocket has been launched can you control its velocity ❹.
There are some things you can always do with the rocket, no matter what
state it is in, and those we place in a generic implementation block ❺. You'll
notice that with the interface designed this way, it's simply not possible for
the user to call a method at the wrong time—we have encoded the usage
rules in the types themselves, and made illegal states *unrepresentable*.

This notion extends to many other domains as well; if your function
ignores a pointer argument unless a given Boolean argument is true, it's
better to combine the two arguments instead. With an enum type with
one variant for false (and no pointer) and one variant for true that holds a
pointer, neither the caller nor the implementer can misunderstand the relationship between the two. This is a powerful idea that I highly encourage
you to make use of.

Another small but useful tool in making interfaces obvious is the #[must
_use] annotation. Add it to any type, trait, or function, and the compiler will
issue a warning if the user's code receives an element of that type or trait, or
calls that function, and does not explicitly handle it. You may already have
seen this in the context of Result: if a function returns a Result and you do
not assign its return value somewhere, you get a compiler warning. Be careful not to overuse this annotation, though—add it only if the user is very
likely to make a mistake if they are not using the return value.

Constrained

Over time, some user will depend on every property of your interface,
whether bug or feature. This is especially true for publicly available libraries where you have no control over your users. As a result, you should think
carefully before you make user-visible changes. Whether you're adding a
new type, field, method, or trait implementation or changing an existing
one, you want to make sure that the change will not break existing users'
code, and that you are planning to keep that change around for a while.
Frequent backward incompatible changes (major version increases in
semantic versioning) are sure to draw the ire of your users.

Many backward incompatible changes are obvious, like renaming a
public type or removing a public method, but some are subtler and tie
in deeply with the way Rust works. Here, we'll cover some of the thornier
subtle changes and how to plan for them. You'll see that you need to balance some of these against how flexible you want your interface to be—
sometimes, something's got to give.

Type Modifications

Removing or renaming a public type will almost certainly break some user's
code. To counter this, you'll want to take advantage of Rust's visibility modifiers, like pub(crate) and pub(in path), whenever possible. The fewer public
types you have, the more freedom you have to change things later without
breaking existing code.

User code can depend on your types in more ways than just by name, though. Consider the public type in Listing 3-3 and the given use of that code.

```
// in your interface
pub struct Unit;
// in user code
let u = lib::Unit;
```

Listing 3-3: An innocent-looking public type

Now consider what happens if you add a private field to Unit. Even though the field you add is private, the change will still break the user's code, because the constructor they relied on has disappeared. Similarly, consider the code and use in Listing 3-4.

```
// in your interface
pub struct Unit { pub field: bool };
// in user code
fn is_true(u: lib::Unit) -> bool {
    matches!(u, Unit { field: true })
}
```

Listing 3-4: User code accessing a single public field

Here, too, adding a private field to Unit will break user code, this time because Rust's exhaustive pattern match checking logic is able to see parts of the interface that the user cannot see. It recognizes that there are more fields, even though the user code cannot access them, and rejects the user's pattern as incomplete. A similar issue arises if we turn a tuple struct into a regular struct with named fields: even if the fields themselves are exactly the same, any old patterns will no longer be valid for the new type definition.

Rust provides the #[non_exhaustive] attribute to help mitigate these issues. You can add it to any type definition, and the compiler will disallow the use of implicit constructors (like lib::Unit { field1: true }) and non-exhaustive pattern matches (that is, patterns without a trailing , ..) on that type. This is a great attribute to add if you suspect that you're likely to modify a particular type in the future. It does constrain user code though, such as by taking away users' ability to rely on exhaustive pattern matches, so avoid adding it if you think a given type is likely to remain stable.

Trait Implementations

As you'll recall from Chapter 2, Rust's coherence rules disallow multiple implementations of a given trait for a given type. Since we do not know what implementations downstream code may have added, adding a blanket implementation of an existing trait is generally a breaking change. The same holds true for implementing a foreign trait for an existing type, or an existing trait for a foreign type—in both cases, the owner of the foreign trait or type may simultaneously add a conflicting implementation, so this must be a breaking change.

Removing a trait implementation is a breaking change, but implementing traits for a *new* type is never a problem, since no crate can have implementations that conflict with that type.

Perhaps counterintuitively, you also want to be careful about implementing *any* trait for an existing type. To see why, consider the code in Listing 3-5.

```
// crate1 1.0
pub struct Unit;
put trait Foo1 { fn foo(&self) }
// note that Foo1 is not implemented for Unit

// crate2; depends on crate1 1.0
use crate1::{Unit, Foo1};
trait Foo2 { fn foo(&self) }
impl Foo2 for Unit { .. }
fn main() {
  Unit.foo();
}
```

Listing 3-5: Implementing a trait for an existing type may cause problems.

If you add impl Foo1 for Unit to crate1 without marking it a breaking change, the downstream code will suddenly stop compiling since the call to foo is now ambiguous. This can even apply to implementations of *new* public traits, if the downstream crate uses wildcard imports (use crate1::*). You will particularly want to keep this in mind if you provide a prelude module that you instruct users to use wildcard imports for.

Most changes to existing traits are also breaking changes, such as changing a method signature or adding a new method. Changing a method signature breaks all implementations, and probably many uses, of the trait, whereas adding a new method "just" breaks all implementations. Adding a new method with a default implementation is fine though, since existing implementations will continue to apply.

I say "generally" and "most" here, because as interface authors, we have a tool available to us that lets us skirt some of these rules: *sealed traits*. A sealed trait is one that can be used only, and not implemented, by other crates. This immediately makes a number of breaking changes non-breaking. For example, you can add a new method to a sealed trait, since you know there are no implementations outside of the current crate to consider. Similarly, you can implement a sealed trait for new foreign types, since you know the foreign crate that defined that type cannot have added a conflicting implementation.

Sealed traits are most commonly used for *derived* traits—traits that provide blanket implementations for types that implement particular other traits. You should seal a trait only if it does not make sense for a foreign crate to implement your trait; it severely restricts the usefulness of the trait, since downstream crates will no longer be able to implement it for their own types. You can also use sealed traits to restrict which types can be used as type arguments, such as restricting the Stage type in the Rocket example from Listing 3-2 to only the Grounded and Launched types.

Listing 3-6 shows how to seal a trait and how to then still add implementations for it in the defining crate.

```
pub trait CanUseCannotImplement: sealed::Sealed ❶ { .. }
mod sealed {
   pub trait Sealed {}
❷ impl<T> Sealed for T where T: TraitBounds {}
}
impl<T> CanUseCannotImplement for T where T: TraitBounds {}
```

Listing 3-6: How to seal a trait and add implementations for it

The trick is to add a private, empty trait as a supertrait of the trait you wish to seal ❶. Since the supertrait is in a private module, other crates cannot reach it and thus cannot implement it. The sealed trait requires the underlying type to implement Sealed, so only the types that we explicitly allow ❷ are able to ultimately implement the trait.

NOTE *If you do seal a trait this way, make sure you document that fact so that users do not get frustrated trying to implement the trait themselves!*

Hidden Contracts

Sometimes, changes you make to one part of your code affect the contract elsewhere in your interface in subtle ways. The two primary ways this happens are through re-exports and auto-traits.

Re-Exports

If any part of your interface exposes foreign types, then any change to one of those foreign types is *also* a change to your interface. For example, consider what happens if you move to a new major version of a dependency and expose a type from that dependency as, say, an iterator type in your interface. A user that depends on your interface may also depend directly on that dependency and expect that the type your interface provides is the same as the one by the same name in that dependency. But if you change the major version of your dependency, that is no longer true even though the *name* of the type is the same. Listing 3-7 shows an example of this.

```
// your crate: bestiter
pub fn iter<T>() -> itercrate::Empty<T> { .. }
// their crate
struct EmptyIterator { it: itercrate::Empty<()> }
EmptyIterator { it: bestiter::iter() }
```

Listing 3-7: Re-exports make foreign crates part of your interface contract.

If your crate moves from itercrate 1.0 to itercrate 2.0 but otherwise does not change, the code in this listing will no longer compile. Even though no types have changed, the compiler believes (correctly) that itercrate1.0::Empty and itercrate2.0::Empty are *different* types. Therefore, you cannot assign the latter to the former, making this a breaking change in your interface.

To mitigate issues like this, it's often best to wrap foreign types using the newtype pattern, and then expose only the parts of the foreign type that you think are useful. In many cases, you can avoid the newtype wrapper altogether by using impl Trait to provide only the very minimal contract to the caller. By promising less, you make fewer changes breaking.

THE SEMVER TRICK

The itercrate example may have rubbed you the wrong way. If the Empty type did not change, then why does the compiler not allow anything that uses it to keep working, regardless of whether the code is using version 1.0 or 2.0 of it? The answer is . . . complicated. It boils down to the fact that the Rust compiler does not assume that just because two types have the same fields, they are the same. To take a simple example of this, imagine that itercrate 2.0 added a #[derive(Copy)] for Empty. Now, the type suddenly has different move semantics depending on whether you are using 1.0 or 2.0! And code written with one in mind won't work with the other.

This problem tends to crop up in large, widely used libraries, where over time, breaking changes are likely to have to happen *somewhere* in the crate. Unfortunately, semantic versioning happens at the crate level, not the type level, so a breaking change anywhere is a breaking change everywhere.

But all is not lost. A few years ago, David Tolnay (the author of serde, among a vast number of other Rust contributions) came up with a neat trick to handle exactly this kind of situation. He called it "the semver trick." The idea is simple: if some type T stays the same across a breaking change (from 1.0 to 2.0, say), then after releasing 2.0, you can release a new 1.0 *minor* version that depends on 2.0 and replaces T with a re-export of T from 2.0.

By doing this, you're ensuring that there is in fact only a single type T across both major versions. This, in turn, means that any crate that depends on 1.0 will be able to use a T from 2.0, and vice versa. And because this happens only for types you *explicitly* opt into with this trick, changes that were in fact breaking will continue to be.

Auto-Traits

Rust has a handful of traits that are automatically implemented for every type depending on what that type contains. The most relevant of these for this discussion are Send and Sync, though the Unpin, Sized, and UnwindSafe traits have similar issues. By their very nature, these add a hidden promise made by nearly every type in your interface. These traits even propagate through otherwise type-erased types like impl Trait.

Implementations for these traits are (generally) automatically added by the compiler, but that also means that they are *not* automatically added if they no longer apply. So, if you have a public type A that contains a private type B, and you change B so that it is no longer Send, then A is now *also* not Send. That is a breaking change!

These changes can be hard to keep track of and are often not discovered until a user of your interface complains that their code no longer works. To catch these cases before they happen, it's good practice to include some simple tests in your test suite that check that all your types implement these traits the way you expect. Listing 3-8 gives an example of what such a test might look like.

```
fn is_normal<T: Sized + Send + Sync + Unpin>() {}
#[test]
fn normal_types() {
  is_normal::<MyType>();
}
```

Listing 3-8: Testing that a type implements a set of traits

Notice that this test does not run any code, but simply tests that the code compiles. If MyType no longer implements Sync, the test code will not compile, and you will know that the change you just made broke the auto-trait implementation.

HIDING ITEMS FROM DOCUMENTATION

The #[doc(hidden)] attribute lets you hide a public item from your documentation without making it inaccessible to code that happens to know it is there. This is often used to expose methods and types that are needed by macros, but not by user code. How such hidden items interact with your interface contract is a matter of some debate. In general, items marked as #[doc(hidden)] are only considered part of your contract insofar as their public effects; for example, if user code may end up containing a hidden type, then whether that type is Send or not is part of the contract, whereas its name is not. Hidden inherent methods and hidden trait methods on sealed traits are not generally part of your interface contract, though you should make sure to state this clearly in the documentation for those methods. And yes, hidden items should still be documented!

Summary

In this chapter we've explored the many facets of designing a Rust interface, whether it's intended for external use or just as an abstraction boundary between the different modules within your crate. We covered a lot of specific pitfalls and tricks, but ultimately, the high-level principles are what should guide your thinking: your interfaces should be unsurprising, flexible, obvious, and constrained. In the next chapter, we will dig into how to represent and handle errors in Rust code.

4

ERROR HANDLING

For all but the simplest programs, you will have methods that can fail. In this chapter, we'll look at different ways to represent, handle, and propagate those failures and the advantages and drawbacks of each. We'll start by exploring different ways to represent errors, including enumeration and erasure, and then examine some special error cases that require a different representation technique. Next, we'll look at various ways of handling errors and the future of error handling.

It's worth noting that best practices for error handling in Rust are still an active topic of conversation, and at the time of writing, the ecosystem has not yet settled on a single, unified approach. This chapter will therefore focus on the underlying principles and techniques rather than recommending specific crates or patterns.

Representing Errors

When you write code that can fail, the most important question to ask yourself is how your users will interact with any errors returned. Will users need to know exactly which error happened and the minutiae about what went wrong, or will they simply log that an error occurred and move on as best they can? To understand this, we have to look at whether the nature of the error is likely to affect what the caller does upon receiving it. This in turn will dictate how we represent different errors.

You have two main options for representing errors: enumeration and erasure. That is, you can either have your error type *enumerate* the possible error conditions so that the caller can distinguish them, or you can just provide the caller with a single, *opaque* error. Let's discuss these two options in turn.

Enumeration

For our example, we'll use a library function that copies bytes from some input stream into some output stream, much like `std::io::copy`. The user provides you with two streams, one to read from and one to write to, and you copy the bytes from one to the other. During this process, it's entirely possible for either stream to fail, at which point the copy has to stop and return an error to the user. Here, the user will likely want to know whether it was the input stream or the output stream that failed. For example, in a web server, if an error occurs on the input stream while streaming a file to a client, it might be because a disk was ejected, whereas if the output stream errors, maybe the client just disconnected. The latter may be an error the server should ignore, since copies to new connections can still complete, whereas the former may require that the whole server be shut down!

This is a case where we want to enumerate the errors. The user needs to be able to distinguish between the different error cases so that they can respond appropriately, so we use an `enum` named `CopyError`, with each variant representing a separate underlying cause for the error, like in Listing 4-1.

```
pub enum CopyError {
  In(std::io::Error),
  Out(std::io::Error),
}
```

Listing 4-1: An enumerated error type

Each variant also includes the error that was encountered to provide the caller with as much information about went wrong as possible.

When making your own error type, you need to take a number of steps to make the error type play nicely with the rest of the Rust ecosystem. First, your error type should implement the `std::error::Error` trait, which provides callers with common methods for introspecting error types. The main method of interest is `Error::source`, which provides a mechanism to find the underlying cause of an error. This is most commonly used to print a backtrace that displays a trace all the way back to the error's root cause. For our `CopyError` type, the implementation of source is straightforward: we match on self and extract and return the inner std::io::Error.

Second, your type should implement both Display and Debug so that callers can meaningfully print your error. This is required if you implement the Error trait. In general, your implementation of Display should give a one-line description of what went wrong that can easily be folded into other error messages. The display format should be lowercase and without trailing punctuation so that it fits nicely into other, larger error reports. Debug should provide a more descriptive error including auxiliary information that may be useful in tracking down the cause of the error, such as port numbers, request identifiers, filepaths, and the like, which #[derive(Debug)] is usually sufficient for.

NOTE   *In older Rust code, you may see references to the Error::description method, but this has been deprecated in favor of Display.*

Third, your type should, if possible, implement both Send and Sync so that users are able to share the error across thread boundaries. If your error type is not thread-safe, you will find that it's almost impossible to use your crate in a multithreaded context. Error types that implement Send and Sync are also much easier to use with the very common std::io::Error type, which is able to wrap errors that implement Error, Send, and Sync. Of course, not all error types can reasonably be Send and Sync, such as if they're tied to particular thread-local resources, and that's okay. You're probably not sending those errors across thread boundaries either. However, it's something to be aware of before you go placing Rc<String> and RefCell<bool> types in your errors.

Finally, where possible, your error type should be 'static. The most immediate benefit of this is that it allows the caller to more easily propagate your error up the call stack without running into lifetime issues. It also enables your error type to be used more easily with type-erased error types, as we'll see shortly.

Opaque Errors

Now let's consider a different example: an image decoding library. You give the library a bunch of bytes to decode, and it gives you access to various image manipulation methods. If the decoding fails, the user needs to be able to figure out how to resolve the issue, and so must understand the cause. But is it important whether the cause is the size field in the image header being invalid, or the compression algorithm failing to decompress a block? Probably not—the application can't meaningfully recover from either situation, even if it knows the exact cause. In cases like this, you as the library author may instead want to provide a single, opaque error type. This also makes your library a little nicer to use, because there is only one error type in use everywhere. This error type should implement Send, Debug, Display, and Error (including the source method where appropriate), but beyond that, the caller doesn't need to know anything more. You might internally represent more fine-grained error states, but there is no need to expose those to the users of the library. Doing so would only serve to unnecessarily increase the size and complexity of your API.

Exactly what your opaque error type should be is mostly up to you. It could just be a type with all private fields that exposes only limited methods for displaying and introspecting the error, or it could be a severely type-erased error type like Box<dyn Error + Send + Sync + 'static>, which reveals nothing more than the fact that it is an error and does not generally let your users introspect at all. Deciding how opaque to make your error types is mostly a matter of whether there is anything interesting about the error beyond its description. With Box<dyn Error>, you leave your users with little option but to bubble up your error. That might be fine if it truly has no information of value to present to the user—for example, if it's just a dynamic error message or is one of a large number of unrelated errors from deeper inside your program. But if the error has some interesting facets to it, such as a line number or a status code, you may want to expose that through a concrete but opaque type instead.

In general, the community consensus is that errors should be rare and therefore should not add much cost to the "happy path." For that reason, errors are often placed behind a pointer type, such as a Box or Arc. This way, they're unlikely to add much to the size of the overall Result type they're contained within.

One benefit of using type-erased errors is that it allows you to easily combine errors from different sources without having to introduce additional error types. That is, type-erased errors often *compose* nicely, and allow you to express an open-ended set of errors. If you write a function whose return type is Box<dyn Error + ...>, then you can use ? across different error types inside that function, on all sorts of different errors, and they will all be turned into that one common error type.

The 'static bound on Box<dyn Error + Send + Sync + 'static> is worth spending a bit more time on in the context of erasure. I mentioned in the previous section that it's useful for letting the caller propagate the error without worrying about the lifetime bounds of the method that failed, but it serves an even bigger purpose: access to downcasting. *Downcasting* is the process of taking an item of one type and casting it to a more specific type. This is one of the few cases where Rust gives you access to type information at runtime; it's a limited case of the more general type reflection that dynamic languages often provide. In the context of errors, downcasting allows a user to turn a dyn Error into a concrete underlying error type when that dyn Error was originally of that type. For example, the user may want to take a particular action if the error they received was a std::io::Error of kind std::io::ErrorKind ::WouldBlock, but they would not take that same action in any other case. If the user gets a dyn Error, they can use Error::downcast_ref to try to downcast the error into a std::io::Error. The downcast_ref method returns an Option, which tells the user whether or not the downcast succeeded. And here is the key observation: downcast_ref works only if the argument is 'static. If we return an opaque Error that's not 'static, we take away the user's ability to do this kind of error introspection should they wish.

There's some disagreement in the ecosystem about whether a library's type-erased errors (or more generally, its type-erased types) are part of its public and stable API. That is, if the method foo in your library returns lib::MyError as a Box<dyn Error>, would changing foo to return a different error type be a breaking change? The type signature hasn't changed, but users may have written code that assumes that they can use downcast to turn that error back into lib::MyError. My opinion on this matter is that you chose to return Box<dyn Error> (and not lib::MyError) for a reason, and unless explicitly documented, that does not guarantee anything in particular about downcasting.

NOTE *While Box<dyn Error + ...> is an attractive type-erased error type, it counterintuitively does not itself implement Error. Therefore, consider adding your own BoxError type for type erasure in libraries that does implement Error.*

You may wonder how Error::downcast_ref can be safe. That is, how does it know whether a provided dyn Error argument is indeed of the given type T? The standard library even has a trait called Any that is implemented for *any* type, and which implements downcast_ref for dyn Any—how can that be okay? The answer lies in the compiler-supported type std::any::TypeId, which allows you to get a unique identifier for any type. The Error trait has a hidden provided method called type_id, whose default implementation is to return TypeId::of::<Self>(). Similarly, Any has a blanket implementation of impl Any for T, and in that implementation, its type_id returns the same. In the context of these impl blocks, the concrete type of Self is known, so this type_id is the type identifier of the real type. That provides all the information downcast_ref needs. downcast_ref calls self.type_id, which forwards through the vtable for dynamically sized types (see Chapter 2) to the implementation for the underlying type and compares that to the type identifier of the provided downcast type. If they match, then the type behind the dyn Error or dyn Any really is T, and it is safe to cast from a reference to one to a reference to the other.

Special Error Cases

Some functions are fallible but cannot return any meaningful error if they fail. Conceptually, these functions have a return type of Result<T, ()>. In some codebases, you may see this represented as Option<T> instead. While both are legitimate choices for the return type for such a function, they convey different semantic meanings, and you should usually avoid "simplifying" a Result<T, ()> to Option<T>. An Err(()) indicates that an operation failed and should be retried, reported, or otherwise handled exceptionally. None, on the other hand, conveys only that the function has nothing to return; it is usually not considered an exceptional case or something that should be handled. You can see this in the #[must_use] annotation on the Result type—when you get a Result, the language expects that it is important to handle both cases, whereas with an Option, neither case actually needs to be handled.

NOTE
You should also keep in mind that () does not implement the Error trait. This means that it cannot be type-erased into Box<dyn Error> and can be a bit of a pain to use with ?. For this reason, it is often better to define your own unit struct type, implement Error for it, and use that as the error instead of () in these cases.

Some functions, like those that start a continuously running server loop, only ever return errors; unless an error occurs, they run forever. Other functions never error but need to return a Result nonetheless, for example, to match a trait signature. For functions like these, Rust provides the *never type*, written with the ! syntax. The never type represents a value that can never be generated. You cannot construct an instance of this type yourself—the only way to make one is by entering an infinite loop or panicking, or through a handful of other special operations that the compiler knows never return. With Result, when you have an Ok or Err that you know will never be used, you can set it to the ! type. If you write a function that returns Result<T, !>, you will be unable to ever return Err, since the only way to do so is to enter code that will never return. Because the compiler knows that any variant with a ! will never be produced, it can also optimize your code with that in mind, such as by not generating the panic code for an unwrap on Result<T, !>. And when you pattern match, the compiler knows that any variant that contains a ! does not even need to be listed. Pretty neat!

One last curious error case is the error type std::thread::Result. Here's its definition:

```
type Result<T> = Result<T, Box<dyn Any + Send + 'static>>;
```

The error type is type-erased, but it's not erased into a dyn Error as we've seen so far. Instead, it is a dyn Any, which guarantees only that the error is *some* type, and nothing more . . . which is not much of a guarantee at all. The reason for this curious-looking error type is that the error variant of std::thread::Result is produced only in response to a panic; specifically, if you try to join a thread that has panicked. In that case, it's not clear that there's much the joining thread can do other than either ignore the error or panic itself using unwrap. In essence, the error type is "a panic" and the value is "whatever argument was passed to panic!," which can truly be any type (even though it's usually a formatted string).

Propagating Errors

Rust's ? operator acts as a shorthand for *unwrap or return early*, for working easily with errors. But it also has a few other tricks up its sleeve that are worth knowing about. First, ? performs type conversion through the From trait. In a function that returns Result<T, E>, you can use ? on any Result<T, X> where E: From<X>. This is the feature that makes error erasure through Box<dyn Error> so appealing; you can just use ? everywhere and not worry about the particular error type, and it will usually "just work."

FROM AND INTO

The standard library has many conversion traits, but two of the core ones are From and Into. It might strike you as odd to have two: if we have From, why do we need Into, and vice versa? There are a couple of reasons, but let's start with the historical one: it wouldn't have been possible to have just one in the early days of Rust due to the coherence rules discussed in Chapter 2. Or, more specifically, what the coherence rules used to be.

Suppose you want to implement two-way conversion between some local type you have defined in your crate and some type in the standard library. You can write impl<T> From<Vec<T>> for MyType<T> and impl<T> Into<Vec<T>> for MyType<T> easily enough, but if you only had From or Into, you would have to write impl<T> From<MyType<T>> for Vec<T> or impl<T> Into<MyType<T>> for Vec<T>. However, the compiler used to reject those implementations! Only since Rust 1.41.0, when the exception for covered types was added to the coherence rules, are they legal. Before that change, it was necessary to have both traits. And since much Rust code was written before Rust 1.41.0, neither trait can be removed now.

Beyond that historical fact, however, there are also good ergonomic reasons to have both of these traits, even if we could start from scratch today. It is often significantly easier to use one or the other in different situations. For example, if you're writing a method that takes a type that can be turned into a Foo, would you rather write fn(impl Into<Foo>) or fn<T>(T) where Foo: From<T>? And conversely, to turn a string into a syntax identifier, would you rather write Ident::from("foo") or <_ as Into<Ident>>::into("foo")? Both of these traits have their uses, and we're better off having them both.

Given that we do have both, you may wonder which you should use in your code today. The answer, it turns out, is pretty simple: implement From, and use Into in bounds. The reason is that Into has a blanket implementation for any T that implements From, so regardless of whether a type explicitly implements From or Into, it implements Into!

Of course, as simple things frequently go, the story doesn't quite end there. Since the compiler often has to "go through" the blanket implementation when Into is used as a bound, the reasoning for whether a type implements Into is more complicated than whether it implements From. And in some cases, the compiler is not quite smart enough to figure that puzzle out. For this reason, the ? operator at the time of writing uses From, not Into. Most of the time that doesn't make a difference, because most types implement From, but it does mean that error types from old libraries that implement Into instead may not work with ?. As the compiler gets smarter, ? will likely be "upgraded" to use Into, at which point that problem will go away, but it's what we have for now.

The second aspect of ? to be aware of is that this operator is really just syntax sugar for a trait tentatively called Try. At the time of writing, the Try trait has not yet been stabilized, but by the time you read this, it's likely that it, or something very similar, will have been settled on. Since the details haven't all been figured out yet, I'll give you only an outline of how Try works, rather than the full method signatures. At its heart, Try defines a wrapper type whose state is either one where further computation is useful (the happy path), or one where it is not. Some of you will correctly think of monads, though we won't explore that connection here. For example, in the case of Result<T, E>, if you have an Ok(t), you can continue on the happy path by unwrapping the t. If you have an Err(e), on the other hand, you want to stop executing and produce the error value immediately, since further computation is not possible as you don't have the t.

What's interesting about Try is that it applies to more types than just Result. An Option<T>, for example, follows the same pattern—if you have a Some(t), you can continue on the happy path, whereas if you have a None, you want to yield None instead of continuing. This pattern extends to more complex types, like Poll<Result<T, E>>, whose happy path type is Poll<T>, which makes ? apply in far more cases than you might expect. When Try stabilizes, we may see ? start to work with all sorts of types to make our happy path code nicer.

The ? operator is already usable in fallible functions, in doctests, and in fn main. To reach its full potential, though, we also need a way to scope this error handling. For example, consider the function in Listing 4-2.

```
fn do_the_thing() -> Result<(), Error> {
    let thing = Thing::setup()?;
    // .. code that uses thing and ? ..
    thing.cleanup();
    Ok(())
}
```

Listing 4-2: A multi-step fallible function using the ? operator

This won't quite work as expected. Any ? between setup and cleanup will cause an early return from the entire function, which would skip the cleanup code! This is the problem *try blocks* are intended to solve. A try block acts pretty much like a single-iteration loop, where ? uses break instead of return, and the final expression of the block has an implicit break. We can now fix the code in Listing 4-2 to always do cleanup, as shown in Listing 4-3.

```
fn do_the_thing() -> Result<(), Error> {
    let thing = Thing::setup()?;
    let r = try {
        // .. code that uses thing and ? ..
    };
    thing.cleanup();
    r
}
```

Listing 4-3: A multi-step fallible function that always cleans up after itself

Try blocks are also not stable at the time of writing, but there is enough of a consensus on their usefulness that they're likely to land in a form similar to that described here.

Summary

This chapter covered the two primary ways to construct error types in Rust: enumeration and erasure. We looked at when you may want to use each one and the advantages and drawbacks of each. We also took a look at some of the behind-the-scenes aspects of the ? operator and considered how ? may become even more useful going forward. In the next chapter, we'll take a step back from the code and look at how you *structure* a Rust project. We'll look at feature flags, dependency management, and versioning as well as how to manage more complex crates using workspaces and subcrates. See you on the next page!

5

PROJECT STRUCTURE

This chapter provides some ideas for structuring your Rust projects. For simple projects, the structure set up by cargo new is likely to be something you think little about. You may add some modules to split up the code, and some dependencies for additional functionality, but that's about it. However, as a project grows in size and complexity, you'll find that you need to go beyond that. Maybe the compilation time for your crate is getting out of hand, or you need conditional dependencies, or you need a better strategy for continuous integration. In this chapter, we will look at some of the tools that the Rust language, and Cargo in particular, provide that make it easier to manage such things.

Features

Features are Rust's primary tool for customizing projects. At its core, a feature is just a build flag that crates can pass to their dependencies in order to add optional functionality. Features carry no semantic meaning in and of themselves—instead, *you* choose what a feature means for *your* crate.

Generally, we use features in three ways: to enable optional dependencies, to conditionally include additional components of a crate, and to augment the behavior of the code. Note that all of these uses are *additive*; features can add to the functionality of the crate, but they shouldn't generally do things like remove modules or replace types or function signatures. This stems from the principle that if a developer makes a simple change to their *Cargo.toml*, such as adding a new dependency or enabling a feature, that shouldn't make their crate stop compiling. If a crate has mutually exclusive features, that principle quickly falls by the wayside—if crate A depends on one feature of crate C, and crate B on another mutually exclusive feature of C, adding a dependency on crate B would then break crate A! For that reason, we generally follow the principle that if crate A compiles against crate C with some set of features, it should also compile if all features are enabled on crate C.

Cargo leans into this principle quite hard. For example, if two crates (A and B) both depend on crate C, but they each enable different features on C, Cargo will compile crate C only once, with *all* the features that either A or B requires. That is, it'll take the union of the requested features for C across A and B. Because of this, it's generally hard to add mutually exclusive features to Rust crates; chances are that some two dependents will depend on the crate with different features, and if those features are mutually exclusive, the downstream crate will fail to build.

NOTE *I highly recommend that you configure your continuous integration infrastructure to check that your crate compiles for any combination of its features. One tool that helps you do this is cargo-hack, which you can find at https://github.com/taiki-e/cargo-hack/.*

Defining and Including Features

Features are defined in *Cargo.toml*. Listing 5-1 shows an example of a crate named foo with a simple feature that enables the optional dependency syn.

```
[package]
name = "foo"
...
[features]
derive = ["syn"]

[dependencies]
syn = { version = "1", optional = true }
```

Listing 5-1: A feature that enables an optional dependency

When Cargo compiles this crate, it will not compile the syn crate by default, which reduces compile time (often significantly). The syn crate will be compiled only if a downstream crate needs to use the APIs enabled by the derive feature and explicitly opts in to it. Listing 5-2 shows how such a downstream crate bar would enable the derive feature, and thus include the syn dependency.

```
[package]
name = "bar"
...
```

```
[dependencies]
foo = { version = "1", features = ["derive"] }
```

Listing 5-2: Enabling a feature of a dependency

Some features are used so frequently that it makes more sense to have a crate opt out of them rather than in to them. To support this, Cargo allows you to define a set of default features for a crate. And similarly, it allows you to opt out of the default features of a dependency. Listing 5-3 shows how foo can make its derive feature enabled by default, while also opting out of some of syn's default features and instead enabling only the ones it needs for the derive feature.

```
[package]
name = "foo"
...
[features]
derive = ["syn"]
default = ["derive"]

[dependencies.syn]
version = "1"
default-features = false
features = ["derive", "parsing", "printing"]
optional = true
```

Listing 5-3: Adding and opting out of default features, and thus optional dependencies

Here, if a crate depends on foo and does not explicitly opt out of the default features, it will also compile foo's syn dependency. In turn, syn will be built with only the three listed features, and no others. Opting out of default features this way, and opting in to only what you need, is a great way to cut down on your compile times!

OPTIONAL DEPENDENCIES AS FEATURES

When you define a feature, the list that follows the equal sign is itself a list of features. This might, at first, sound a little odd—in Listing 5-3, syn is a dependency, not a feature. It turns out that Cargo makes every optional dependency a feature with the same name as the dependency. You'll see this if you try to add a feature with the same name as an optional dependency; Cargo won't allow it. Support for a different namespace for features and dependencies is in the works in Cargo, but has not been stabilized at the time of writing. In the meantime, if you want to have a feature named after a dependency, you can rename the dependency using package = "" to avoid the name collision. The list of features that a feature enables can also include features of dependencies. For example, you can write derive = ["syn/derive"] to have your derive feature enable the derive feature of the syn dependency.

Using Features in Your Crate

When using features, you need to make sure your code uses a dependency only if it is available. And if your feature enables a particular component, you need to make sure that if the feature isn't enabled, the component is not included.

You achieve this using *conditional compilation*, which lets you use annotations to give conditions under which a particular piece of code should or should not be compiled. Conditional compilation is primarily expressed using the #[cfg] attribute. There is also the closely related cfg! macro, which lets you change runtime behavior based on similar conditions. You can do all sorts of neat things with conditional compilation, as we'll see later in this chapter, but the most basic form is #[cfg(feature = "some-feature")], which makes it so that the next "thing" in the source code is compiled only if the some-feature feature is enabled. Similarly, if cfg!(feature = "some-feature") is equivalent to if true only if the derive feature is enabled (and if false otherwise).

The #[cfg] attribute is used more often than the cfg! macro, because the macro modifies runtime behavior based on the feature, which can make it difficult to ensure that features are additive. You can place #[cfg] in front of certain Rust *items*—such as functions and type definitions, impl blocks, modules, and use statements—as well as on certain other constructs like struct fields, function arguments, and statements. The #[cfg] attribute can't go just anywhere, though; where it can appear is carefully restricted by the Rust language team so that conditional compilation can't cause situations that are too strange and hard to debug.

Remember that modifying certain public parts of your API may inadvertently make a feature nonadditive, which in turn may make it impossible for some users to compile your crate. You can often use the rules for backward compatible changes as a rule of thumb here—for example, if you make an enum variant or a public struct field conditional upon a feature, then that type must also be annotated with #[non_exhaustive]. Otherwise, a dependent crate that does not have the feature enabled may no longer compile if the feature is added due to some second crate in the dependency tree.

> **NOTE** *If you're writing a large crate where you expect that your users will need only a subset of the functionality, you should consider making it so that larger components (usually modules) are guarded by features. That way, users can opt in to, and pay the compilation cost of, only the parts they really need.*

Workspaces

Crates play many roles in Rust—they are the vertices in the dependency graph, the boundaries for trait coherence, and the scopes for compilation features. Because of this, each crate is managed as a single compilation unit; the Rust compiler treats a crate more or less as one big source file compiled as one chunk that is ultimately turned into a single binary output (either a binary or a library).

While this simplifies many aspects of the compiler, it also means that large crates can be painful to work with. If you change a unit test, a comment, or a type in one part of your application, the compiler must re-evaluate the entire crate to determine what, if anything, changed. Internally, the compiler implements a number of mechanisms to speed up this process, like incremental recompilation and parallel code generation, but ultimately the size of your crate is a big factor in how long your project takes to compile.

For this reason, as your project grows, you may want to split it into multiple crates that internally depend on one another. Cargo has just the feature you need to make this convenient: workspaces. A *workspace* is a collection of crates (often called *subcrates*) that are tied together by a top-level *Cargo.toml* file like the one shown in Listing 5-4.

```
[workspace]
members = [
  "foo",
  "bar/one",
  "bar/two",
]
```

Listing 5-4: A workspace Cargo.toml

The members array is a list of directories that each contain a crate in the workspace. Those crates all have their own *Cargo.toml* files in their own subdirectories, but they share a single *Cargo.lock* file and a single output directory. The crate names don't need to match the entry in members. It is common, but not required, that crates in a workspace share a name prefix, usually chosen as the name of the "main" crate. For example, in the tokio crate, the members are called tokio, tokio-test, tokio-macros, and so on.

Perhaps the most important feature of workspaces is that you can interact with all of the workspace's members by invoking cargo in the root of the workspace. Want to check that they all compile? cargo check will check them all. Want to run all your tests? cargo test will test them all. It's not quite as convenient as having everything in one crate, so don't go splitting everything into minuscule crates, but it's a pretty good approximation.

NOTE *Cargo commands will generally do the "right thing" in a workspace. If you ever need to disambiguate, such as if two workspace crates both have a binary by the same name, use the -p flag (for package). If you are in the subdirectory for a particular workspace crate, you can pass --workspace to perform the command for the entire workspace instead.*

Once you have a workspace-level *Cargo.toml* with the array of workspace members, you can set your crates to depend on one another using path dependencies, as shown in Listing 5-5.

```
# bar/two/Cargo.toml
[dependencies]
one = { path = "../one" }
# bar/one/Cargo.toml
[dependencies]
foo = { path = "../../foo" }
```

Listing 5-5: Intercrate dependencies among workspace crates

Now if you make a change to the crate in *bar/two*, then only that crate
is re-compiled, since foo and *bar/one* did not change. It may even be faster
to compile your project from scratch, since the compiler does not need to
evaluate your entire project source for optimization opportunities.

SPECIFYING INTRA-WORKSPACE DEPENDENCIES

The most obvious way to specify that one crate in a workspace depends on
another is to use the path specifier, as shown in Listing 5-5. However, if your
individual subcrates are intended for public consumption, you may want to use
version specifiers instead.

Say you have a crate that depends on a Git version of the one crate from
the bar workspace in Listing 5-5 with one = { git = ". . ." }, and a released
version of foo (also from bar) with foo = "1.0.0". Cargo will dutifully fetch
the one Git repository, which holds the entire bar workspace, and see that one
in turn depends on foo, located at *../../foo* inside the workspace. But Cargo
doesn't know that the released version foo = "1.0.0" and the foo in the Git
repository are the same crate! It considers them two separate dependencies
that just happen to have the same name.

You may already see where this is going. If you try to use any type from
foo (1.0.0) with an API from one that accepts a type from foo, the compiler will
reject the code. Even though the types have the same name, the compiler can't
know that they are the same underlying type. And the user will be thoroughly
confused, since the compiler will say something like "expected foo::Type, got
foo::Type."

The best way to mitigate this problem is to use path dependencies between
subcrates only if they depend on unpublished changes. As long as one works
with foo 1.0.0, it should list foo = "1.0.0" in its dependencies. Only if you
make a change to foo that one needs should you change one to use a path
dependency. And once you release a new version of foo that one can depend
on, you should remove the path dependency again.

This approach also has its shortcomings. Now if you change foo and then
run the tests for one, you'll see that one will be tested using the old foo, which
may not be what you expected. You'll probably want to configure your continuous integration infrastructure to test each subcrate both with the latest released
versions of the other subcrates and with all of them configured to use path
dependencies.

Project Configuration

Running `cargo new` sets you up with a minimal *Cargo.toml* that has the crate's name, its version number, some author information, and an empty list of dependencies. That will take you pretty far, but as your project matures, there are a number of useful things you may want to add to your *Cargo.toml*.

Crate Metadata

The first and most obvious thing to add to your *Cargo.toml* is all the metadata directives that Cargo supports. In addition to obvious fields like `description` and `homepage`, it can be useful to include information such as the path to a *README* for the crate (`readme`), the default binary to run with `cargo run` (`default-run`), and additional `keywords` and `categories` to help *crates.io* categorize your crate.

For crates with a more convoluted project layout, it's also useful to set the `include` and `exclude` metadata fields. These dictate which files should be included and published in your package. By default, Cargo includes all files in a crate's directory except any listed in your *.gitignore* file, but this may not be what you want if you also have large test fixtures, unrelated scripts, or other auxiliary data in the same directory that you *do* want under version control. As their names suggest, `include` and `exclude` allow you to include only a specific set of files or exclude files matching a given set of patterns, respectively.

NOTE *If you have a crate that should never be published, or should be published only to certain alternative registries (that is, not to* crates.io*), you can set the* `publish` *directive to* `false` *or to a list of allowed registries.*

The list of metadata directives you can use continues to grow, so make sure to periodically check in on the Manifest Format page of the Cargo reference (*https://doc.rust-lang.org/cargo/reference/manifest.html*).

Build Configuration

Cargo.toml can also give you control over how Cargo builds your crate. The most obvious tool for this is the `build` parameter, which allows you to write a completely custom build program for your crate (we'll revisit this in Chapter 11). However, Cargo also provides two smaller, but very useful, mechanisms that we'll explore here: patches and profiles.

[patch]

The [patch] section of *Cargo.toml* allows you to specify a different source for a dependency that you can use temporarily, no matter where in your dependencies the patched dependency appears. This is invaluable when you need to compile your crate against a modified version of some transitive dependency to test a bug fix, a performance improvement, or a new minor release you're about to publish. Listing 5-6 shows an example of how you might temporarily use a variant of a set of dependencies.

```
[patch.crates-io]
# use a local (presumably modified) source
regex = { path = "/home/jon/regex" }
# use a modification on a git branch
serde = { git = "https://github.com/serde-rs/serde.git", branch = "faster" }
# patch a git dependency
[patch.'https://github.com/jonhoo/project.git']
project = { path = "/home/jon/project" }
```

Listing 5-6: Overriding dependency sources in Cargo.toml *using [patch]*

Even if you patch a dependency, Cargo takes care to check the crate versions so that you don't accidentally end up patching the wrong major version of a crate. If you for some reason transitively depend on multiple major versions of the same crate, you can patch each one by giving them distinct identifiers, as shown in Listing 5-7.

```
[patch.crates-io]
nom4 = { path = "/home/jon/nom4", package = "nom" }
nom5 = { path = "/home/jon/nom5", package = "nom" }
```

Listing 5-7: Overriding multiple versions of the same crate in Cargo.toml *using [patch]*

Cargo will look at the *Cargo.toml* inside each path, realize that /nom4 contains major version 4 and that /nom5 contains major version 5, and patch the two versions appropriately. The package keyword tells Cargo to look for a crate by the name nom in both cases instead of using the dependency identifiers (the part on the left) as it does by default. You can use package this way in your regular dependencies as well to rename a dependency!

Keep in mind that patches are not taken into account in the package that's uploaded when you publish a crate. A crate that depends on your crate will use only its own [patch] section (which may be empty), not that of your crate!

CRATES VS. PACKAGES

You may wonder what the difference between a package and a crate is. These two terms are often used interchangeably in informal contexts, but they also have specific definitions that vary depending on whether you're talking about the Rust compiler, Cargo, *crates.io*, or something else. I personally think of a crate as a Rust module hierarchy starting at a root *.rs* file (one where you can use crate-level attributes like #![feature])—usually something like *lib.rs* or *main.rs*. In contrast, a package is a collection of crates and metadata, so essentially all that's described by a *Cargo.toml* file. That may include a library crate, multiple binary crates, some integration test crates, and maybe even multiple workspace members that themselves have *Cargo.toml* files.

[profile]

The [profile] section lets you pass additional options to the Rust compiler in order to change the way it compiles your crate. These options fall primarily into three categories: performance options, debugging options, and options that change code behavior in user-defined ways. They all have different defaults depending on whether you are compiling in debug mode or in release mode (other modes also exist).

The three primary performance options are opt-level, codegen-units, and lto. The opt-level option tweaks runtime performance by telling the compiler how aggressively to optimize your program (0 is "not at all," 3 is "as much as you can"). The higher the setting, the more optimized your code will be, which *may* make it run faster. Extra optimization comes at the cost of higher compile times, though, which is why optimizations are generally enabled only for release builds.

NOTE *You can also set opt-level to "s" to optimize for binary size, which may be important on embedded platforms.*

The codegen-units option is about compile-time performance. It tells the compiler how many independent compilation tasks (*code generation units*) it is allowed to split the compilation of a single crate into. The more pieces a large crate's compilation is split into, the faster it will compile, since more threads can help compile the crate in parallel. Unfortunately, to achieve this speedup, the threads need to work more or less independently, which means code optimization suffers. Imagine, for example, that the segment of a crate compiling in one thread could benefit from inlining some code in a different segment—since the two segments are independent, that inlining can't happen! This setting, then, is a trade-off between compile-time performance and runtime performance. By default, Rust uses an effectively unbounded number of codegen units in debug mode (basically, "compile as fast as you can") and a smaller number (16 at the time of writing) in release mode.

The lto setting toggles *link-time optimization (LTO)*, which enables the compiler (or the linker, if you want to get technical about it) to jointly optimize bits of your program, known as *compilation units*, that were originally compiled separately. The exact details of LTO are beyond the scope of this book, but the basic idea is that the output from each compilation unit includes information about the code that went into that unit. After all the units have been compiled, the linker makes another pass over all of the units and uses that additional information to optimize the combined compiled code. This extra pass adds to the compile time but recovers most of the runtime performance that may have been lost due to splitting the compilation into smaller parts. In particular, LTO can offer significant performance boosts to performance-sensitive programs that might benefit from cross-crate optimization. Beware, though, that cross-crate LTO can add a lot to your compile time.

Rust performs LTO across all the codegen units within each crate by default in an attempt to make up for the lost optimizations caused by using many codegen units. Since the LTO is performed only within each crate, rather than across crates, this extra pass isn't too onerous, and the added compile time should be lower than the amount of time saved by using a lot of codegen units. Rust also offers a technique known as *thin LTO*, which allows the LTO pass to be mostly parallelized, at the cost of missing some optimizations a "full" LTO pass would have found.

NOTE *LTO can be used to optimize across foreign function interface boundaries in many cases, too. See the* `linker-plugin-lto` `rustc` *flag for more details.*

The [profile] section also supports flags that aid in debugging, such as debug, debug-assertions, and overflow-checks. The debug flag tells the compiler to include debug symbols in the compiled binary. This increases the binary size, but it means that you get function names and such, rather than just instruction addresses, in backtraces and profiles. The debug-assertions flag enables the debug_assert! macro and other related debug code that isn't compiled otherwise (through cfg(debug_assertions)). Such code may make your program run slower, but it makes it easier to catch questionable behavior at runtime. The overflow-checks flag, as the name implies, enables overflow checks on integer operations. This slows them down (notice a trend here?) but can help you catch tricky bugs early on. By default, these are all enabled in debug mode and disabled in release mode.

[profile.*.panic]

The [profile] section has another flag that deserves its own subsection: panic. This option dictates what happens when code in your program calls panic!, either directly or indirectly through something like unwrap. You can set panic to either unwind (the default on most platforms) or abort. We'll talk more about panics and unwinding in Chapter 9, but I'll give a quick summary here.

Normally in Rust, when your program panics, the thread that panicked starts *unwinding* its stack. You can think of unwinding as forcibly returning recursively from the current function all the way to the bottom of that thread's stack. That is, if main called foo, foo called bar, and bar called baz, a panic in baz would forcibly return from baz, then bar, then foo, and finally from main, resulting in the program exiting. A thread that unwinds will drop all values on the stack normally, which gives the values a chance to clean up resources, report errors, and so on. This gives the running system a chance to exit gracefully even in the case of a panic.

When a thread panics and unwinds, other threads continue running unaffected. Only when (and if) the thread that ran main exits does the program terminate. That is, the panic is generally isolated to the thread in which the panic occurred.

This means unwinding is a double-edged sword; the program is limping along with some failed components, which may cause all sorts of strange behaviors. For example, imagine a thread that panics halfway through updating the state in a Mutex. Any thread that subsequently acquires that Mutex must now be prepared to handle the fact that the state may be in a partially updated, inconsistent state. For this reason, some synchronization primitives (like Mutex) will remember if a panic occurred when they were last accessed and communicate that to any thread that tries to access the primitive subsequently. If a thread encounters such a state, it will normally also panic, which leads to a cascade that eventually terminates the entire program. But that is arguably better than continuing to run with corrupted state!

The bookkeeping needed to support unwinding is not free, and it often requires special support by the compiler and the target platform. For example, many embedded platforms cannot unwind the stack efficiently at all. Rust therefore supports a different panic mode: abort ensures the whole program simply exits immediately when a panic occurs. In this mode, no threads get to do any cleanup. This may seem severe, and it is, but it ensures that the program is never running in a half-working state and that errors are made visible immediately.

WARNING *The panic setting is global—if you set it to* abort, *all your dependencies are also compiled with* abort.

You may have noticed that when a thread panics, it tends to print a *backtrace*: the trail of function calls that led to where the panic occurred. This is also a form of unwinding, though it is separate from the unwinding panic behavior discussed here. You can have backtraces even with panic=abort by passing -Cforce-unwind-tables to rustc, which makes rustc include the information necessary to walk back up the stack while still terminating the program on a panic.

PROFILE OVERRIDES

You can set profile options for just a particular dependency, or a particular profile, using *profile overrides*. For example, Listing 5-8 shows how to enable aggressive optimizations for the serde crate and moderate optimizations for all other crates in debug mode, using the [profile.*<profile-name>*.package.*<crate-name>*] syntax.

```
[profile.dev.package.serde]
opt-level = 3
[profile.dev.package."*"]
opt-level = 2
```

Listing 5-8: Overriding profile options for a specific dependency or for a specific mode

This kind of optimization override can be handy if some dependency would be prohibitively slow in debug mode (such as decompression or video encoding), and you need it optimized so that your test suite won't take several days to complete. You can also specify global profile defaults using a [profile.dev] (or similar) section in the Cargo configuration file in *~/.cargo/config*.

When you set optimization parameters for a specific dependency, keep in mind that the parameters apply only to the code compiled as part of that crate; if serde in this example has a generic method or type that you use in your crate,

(continued)

> the code of that method or type will be monomorphized and optimized in *your* crate, and *your* crate's profile settings will apply, not those in the profile override for serde.

Conditional Compilation

Most Rust code you write is universal—it'll work the same regardless of what CPU or operating system it runs on. But sometimes you'll have to do something special to get the code to work on Windows, on ARM chips, or when compiled against a particular platform application binary interface (ABI). Or maybe you want to write an optimized version of a particular function when a given CPU instruction is available, or disable some slow but uninteresting setup code when running in a continuous integration (CI) environment. To cater to cases like these, Rust provides mechanisms for *conditional compilation*, in which a particular segment of code is compiled only if certain conditions are true of the compilation environment.

We denote conditional compilation with the cfg keyword that you saw earlier in the chapter in "Using Features in Your Crate." It usually appears in the form of the #[cfg(condition)] attribute, which says to compile the next item only if condition is true. Rust also has #[cfg_attr(condition, attribute)], which is compiled as #[attribute] if condition holds and is a no-op otherwise. You can also evaluate a cfg condition as a Boolean expression using the cfg!(condition) macro.

Every cfg construct takes a single condition made up of options, like feature = "some-feature", and the combinators all, any, and not, which do what you would probably expect. Options are either simple names, like unix, or key/value pairs like those used by feature conditions.

There are a number of interesting options you can make compilation dependent on. Let's go through them, from most common to least common:

Feature options

You've already seen examples of these. Feature options take the form feature = "name-of-feature" and are considered true if the named feature is enabled. You can check for multiple features in a single condition using the combinators. For example, any(feature = "f1", feature = "f2") is true if either feature f1 or feature f2 is enabled.

Operating system options

These use key/value syntax with the key target_os and values like windows, macos, and linux. You can also specify a family of operating systems using target_family, which takes the value windows or unix. These are common enough that they have received their own named short forms, so you can use cfg(windows) and cfg(unix) directly. For example, if you wanted a particular code segment to be compiled only on macOS and Windows, you would write: #[cfg(any(windows, target_os = "macos"))].

Context options

These let you tailor code to a particular compilation context. The most common of these is the test option, which is true only when the crate is being compiled under the test profile. Keep in mind that test is set only for the crate that is being tested, not for any of its dependencies. This also means that test is not set in your crate when running integration tests; it's the integration tests that are compiled under the test profile, whereas your actual crate is compiled normally (that is, without test set). The same applies to the doc and doctest options, which are set only when building documentation or compiling doctests, respectively. There's also the debug_assertions option, which is set in debug mode by default.

Tool options

Some tools, like clippy and Miri, set custom options (more on that later) that let you customize compilation when run under these tools. Usually, these options are named after the tool in question. For example, if you want a particular compute-intensive test not to run under Miri, you can give it the attribute #[cfg_attr(miri, ignore)].

Architecture options

These let you compile based on the CPU instruction set the compiler is targeting. You can specify a particular architecture with target_arch, which takes values like x86, mips, and aarch64, or you can specify a particular platform feature with target_feature, which takes values like avx or sse2. For very low-level code, you may also find the target_endian and target_pointer_width options useful.

Compiler options

These let you adapt your code to the platform ABI it is compiled against and are available through target_env with values like gnu, msvc, and musl. For historical reasons, this value is often empty, especially on GNU platforms. You normally need this option only if you need to interface directly with the environment ABI, such as when linking against an ABI-specific symbol name using #[link].

While cfg conditions are usually used to customize code, some can also be used to customize dependencies. For example, the dependency winrt usually makes sense only on Windows, and the nix crate is probably useful only on Unix-based platforms. Listing 5-9 gives an example of how you can use cfg conditions for this:

```
[target.'cfg(windows)'.dependencies]
winrt = "0.7"
[target.'cfg(unix)'.dependencies]
nix = "0.17"
```

Listing 5-9: Conditional dependencies

Here, we specify that `winrt` version 0.7 should be considered a dependency only under `cfg(windows)` (so, on Windows), and `nix` version 0.17 only under `cfg(unix)` (so, on Linux, macOS, and other Unix-based platforms). One thing to keep in mind is that the `[dependencies]` section is evaluated very early in the build process, when only certain `cfg` options are available. In particular, feature and context options are not yet available at this point, so you cannot use this syntax to pull in dependencies based on features and contexts. You can, however, use any `cfg` that depends only on the target specification or architecture, as well as any options explicitly set by tools that call into `rustc` (like `cfg(miri)`).

NOTE *While we're on the topic of dependency specifications, I highly recommend that you set up your CI infrastructure to perform basic auditing of your dependencies using tools like `cargo-deny` and `cargo-audit`. These tools will detect cases where you transitively depend on multiple major versions of a given dependency, where you depend on crates that are unmaintained or have known security vulnerabilities, or where you use licenses that you may want to avoid. Using such a tool is a great way to raise the quality of your codebase in an automated way!*

It's also quite simple to add your own custom conditional compilation options. You just have to make sure that `--cfg=myoption` is passed to `rustc` when `rustc` compiles your crate. The easiest way to do this is to add your `--cfg` to the `RUSTFLAGS` environment variable. This can come in handy in CI, where you may want to customize your test suite depending on whether it's being run on CI or on a dev machine: add `--cfg=ci` to `RUSTFLAGS` in your CI setup, and then use `cfg(ci)` and `cfg(not(ci))` in your code. Options set this way are also available in *Cargo.toml* dependencies.

Versioning

All Rust crates are versioned and are expected to follow Cargo's implementation of semantic versioning. *Semantic versioning* dictates the rules for what kinds of changes require what kinds of version increases and for which versions are considered compatible, and in what ways. The RFC 1105 standard itself is well worth reading (it's not horribly technical), but to summarize, it differentiates between three kinds of changes: breaking changes, which require a major version change; additions, which require a minor version change; and bug fixes, which require only a patch version change. RFC 1105 does a decent job of outlining what constitutes a breaking change in Rust, and we've touched on some aspects of it elsewhere in this book.

I won't go into detail here about the exact semantics of the different types of changes. Instead, I want to highlight some less straightforward ways version numbers come up in the Rust ecosystem, which you need to keep in mind when deciding how to version your own crates.

Minimum Supported Rust Version

The first Rust-ism is the *minimum supported Rust version (MSRV)*. There is much debate in the Rust community about what policy projects should adhere to when it comes to their MSRV and versioning, and there's no truly good answer. The core of the problem is that some Rust users are limited to using older versions of Rust, often in an enterprise setting where they have little choice. If we constantly take advantage of newly stabilized APIs, those users will not be able to compile the latest versions of our crates and will be left behind.

There are two techniques crate authors can use to make life a little easier for users in this position. The first is to establish an MSRV policy promising that new versions of a crate will always compile with any stable release from the last *X* months. The exact number varies, but 6 or 12 months is common. With Rust's six-week release cycle, that corresponds to the latest four or eight stable releases, respectively. Any new code introduced to the project must compile with the MSRV compiler (usually checked by CI) or be held until the MSRV policy allows it to be merged as is. This can sometimes be a pain, as it means these crates cannot take advantage of the latest and greatest the language has to offer, but it will make life easier for your users.

The second technique is to make sure to increase the minor version number of your crate any time that the MSRV changes. So, if you release version 2.7.0 of your crate and that increases your MSRV from Rust 1.44 to Rust 1.45, then a project that is stuck on 1.44 and that depends on your crate can use the dependency version specifier version = "2, <2.7" to keep the project working until it can move on to Rust 1.45. It's important that you increment the minor version, not just the patch version, so that you can still issue critical security fixes for the previous MSRV release by doing another patch release if necessary.

Some projects take their MSRV support so seriously that they consider an MSRV change a breaking change and increment the major version number. This means that downstream projects will explicitly have to opt in to an MSRV change, rather than opting out—but it also means that users who do not have such strict MSRV requirements will not see future bug fixes without updating their dependencies, which may require *them* to issue a breaking change as well. As I said, none of these solutions are without drawbacks.

Enforcing an MSRV in the Rust ecosystem today is challenging. Only a small subset of crates provide any MSRV guarantees, and even if your dependencies do, you will need to constantly monitor them to know when they increase their MSRV. When they do, you'll need to do a new release of your crate with the restricted version bounds mentioned previously to make sure your MSRV doesn't also change. This may in turn force you to forego security and performance updates made to your dependencies, as you'll have to continue using older versions until your MSRV policy permits updating. And that decision also carries over to your dependents. There have been proposals to build MSRV checking into Cargo itself, but nothing workable has been stabilized as of this writing.

Minimal Dependency Versions

When you first add a dependency, it's not always clear what version specifier you should give that dependency. Programmers commonly choose the latest version, or just the current major version, but chances are that both of those choices are wrong. By "wrong," I don't mean that your crate won't compile, but rather that making that choice may cause strife for users of your crate down the line. Let's look at why each of these cases is problematic.

First, consider the case where you add a dependency on hugs = "1.7.3", the latest published version. Now imagine that a developer somewhere depends on your crate, but they also depend on some other crate, foo, that itself depends on hugs. Further imagine that the author of foo is really careful about their MSRV policy, so they depend on hugs = "1, <1.6". Here, you'll run into trouble. When Cargo sees hugs = "1.7.3", it considers only versions >=1.7. But then it sees that foo's dependency on hugs requires <1.6, so it gives up and reports that there is no version of hugs compatible with all the requirements.

NOTE *In practice, there are a number of reasons why a crate may explicitly not want a newer version of a dependency. The most common ones are to enforce MSRV, to meet enterprise auditing requirements (the newer version will contain code that hasn't been audited), and to ensure reproducible builds where only the exact listed version is used.*

This is unfortunate, as it could well be that your crate compiles fine with, say, hugs 1.5.6. Maybe it even compiles fine with *any* 1.X version! But by using the latest version number, you are telling Cargo to consider only versions at or beyond that minor version. Is the solution to use hugs = "1" instead, then? No, that's not quite right either. It could be that your code truly does depend on something that was added only in hugs 1.6, so while 1.6.2 would be fine, 1.5.6 would not be. You wouldn't notice this if you were only ever compiling your crate in situations where a newer version ends up getting used, but if some crate in the dependency graph specifies hugs = "1, <1.5", your crate would not compile!

The right strategy is to list the earliest version that has all the things your crate depends on and to make sure that this remains the case even as you add new code to your crate. But how do you establish that beyond trawling the changelogs, or through trial and error? Your best bet is to use Cargo's unstable -Zminimal-versions flag, which makes your crate use the minimum acceptable version for all dependencies, rather than the maximum. Then, set all your dependencies to just the latest major version number, try to compile, and add a minor version to any dependencies that don't. Rinse and repeat until everything compiles fine, and you now have your minimum version requirements!

It's worth noting that, like with MSRV, minimal version checking faces an ecosystem adoption problem. While you may have set all your version specifiers correctly, the projects you depend on may not have. This makes the Cargo minimal versions flag hard to use in practice (and is why it's still unstable). If you depend on foo, and foo depends on bar with a specifier of bar = "1" when it actually requires bar = "1.4", Cargo will report that it failed to compile foo no matter how you list foo because the -Z flag tells it to always prefer minimal versions. You can work around this by listing bar directly in *your* dependencies with the appropriate version requirement, but these workarounds can be painful to set up and maintain. You may end up listing a large number of dependencies that are only really pulled in through your transitive dependencies, and you'll have to keep that list up to date as time goes on.

NOTE *One current proposal is to present a flag that favors minimal versions for the current crate but maximal ones for dependencies, which seems quite promising.*

Changelogs

For all but the most trivial crates, I highly recommend keeping a changelog. There is little more frustrating than seeing that a dependency has received a major version bump and then having to dig through the Git logs to figure out what changed and how to update your code. I recommend that you do not just dump your Git logs into a file named *changelog*, but instead keep a manual changelog. It is much more likely to be useful.

A simple but good format for changelogs is the Keep a Changelog format documented at *https://keepachangelog.com/*.

Unreleased Versions

Rust considers version numbers even when the source of a dependency is a directory or a Git repository. This means that semantic versioning is important even when you have not yet published a release to *crates.io*; it matters what version is listed in your *Cargo.toml* between releases. The semantic versioning standard does not dictate how to handle this case, but I'll provide a workflow that works decently well without being too onerous.

After you've published a release, immediately update the version number in your *Cargo.toml* to the next patch version with a suffix like *-alpha.1*. If you just released 2.0.3, make the new version 2.0.4-alpha.1. If you just released an alpha, increment the alpha number instead.

As you make changes to the code between releases, keep an eye out for additive or breaking changes. If one happens, and the corresponding version number has not changed since the last release, increment it. For example, if the last released version is 2.0.3, the current version is 2.0.4-alpha.2, and you make an additive change, make the version with the change 2.1.0-alpha.1. If you made a breaking change, it becomes 3.0.0-alpha.1 instead. If the corresponding version increase has already been made, just increment the alpha number.

When you make a release, remove the suffix (unless you want to do a prerelease), then publish, and start from the top.

This process is effective because it makes two common workflows work much better. First, imagine that a developer depends on major version 2 of your crate, but they need a feature that's currently available only in Git. Then you commit a breaking change. If you don't increase the major version at the same time, their code will suddenly fail in unexpected ways, either by failing to compile or as a result of weird runtime issues. If you follow the procedure laid out here, they'll instead be notified by Cargo that a breaking change has occurred, and they'll have to either resolve that or pin a specific commit.

Next, imagine that a developer needs a feature they just contributed to your crate, but which isn't part of any released version of your crate yet. They've used your crate behind a Git dependency for a while, so other developers on their project already have older checkouts of your crate's repository. If you do not increment the major version number in Git, this developer has no way to communicate that their project now relies on the feature that was just merged. If they push their change, their fellow developers will find that the project no longer compiles, since Cargo will reuse the old checkout. If, on the other hand, the developer can increment the minor version number for the Git dependency, then Cargo will realize that the old checkout is outdated.

This workflow is by no means perfect. It doesn't provide a good way to communicate multiple minor or major changes between releases, and you still need to do a bit of work to keep track of the versions. However, it does address two of the most common issues Rust developers run into when they work against Git dependencies, and even if you make multiple such changes between releases, this workflow will still catch many of the issues.

If you're not too worried about small or consecutive version numbers in releases, you can improve this suggested workflow by simply always incrementing the appropriate part of the version number. Be aware, though, that depending on how frequently you make such changes, this may make your version numbers quite large!

Summary

In this chapter, we've looked at a number of mechanisms for configuring, organizing, and publishing crates, for both your own benefit and that of others. We've also gone over some common gotchas when working with dependencies and features in Cargo that now hopefully won't catch you out in the future. In the next chapter we'll turn to testing and dig into how you go beyond Rust's simple #[test] functions that we know and love.

6

TESTING

In this chapter, we'll look at the various ways in which you can extend Rust's testing capabilities and what other kinds of testing you may want to add into your testing mix. Rust comes with a number of built-in testing facilities that are well covered in *The Rust Programming Language*, represented primarily by the #[test] attribute and the *tests/* directory. These will serve you well across a wide range of applications and scales and are often all you need when you are getting started with a project. However, as the codebase develops and your testing needs grow more elaborate, you may need to go beyond just tagging #[test] onto individual functions.

This chapter is divided into two main sections. The first part covers Rust testing mechanisms, like the standard testing harness and conditional testing code. The second looks at other ways to evaluate the correctness of your Rust code, such as benchmarking, linting, and fuzzing.

Rust Testing Mechanisms

To understand the various testing mechanisms Rust provides, you must first understand how Rust builds and runs tests. When you run `cargo test --lib`, the only special thing Cargo does is pass the `--test` flag to `rustc`. This flag tells `rustc` to produce a test binary that runs all the unit tests, rather than just compiling the crate's library or binary. Behind the scenes, `--test` has two primary effects. First, it enables `cfg(test)` so that you can conditionally include testing code (more on that in a bit). Second, it makes the compiler generate a *test harness*: a carefully generated `main` function that invokes each `#[test]` function in your program when it's run.

The Test Harness

The compiler generates the test harness `main` function through a mix of procedural macros, which we'll discuss in greater depth in Chapter 7, and a light sprinkling of magic. Essentially, the harness transforms every function annotated by `#[test]` into a test *descriptor*—this is the procedural macro part. It then exposes the path of each of the descriptors to the generated `main` function—this is the magic part. The descriptor includes information like the test's name, any additional options it has set (like `#[should_panic]`), and so on. At its core, the test harness iterates over the tests in the crate, runs them, captures their results, and prints the results. So, it also includes logic to parse command line arguments (for things like `--test-threads=1`), capture test output, run the listed tests in parallel, and collect test results.

As of this writing, Rust developers are working on making the magic part of test harness generation a publicly available API so that developers can build their own test harnesses. This work is still at the experimental stage, but the proposal aligns fairly closely with the model as it exists today. Part of the magic that needs to be figured out is how to ensure that `#[test]` functions are available to the generated `main` function even if they are inside private submodules.

Integration tests (the tests in *tests/*) follow the same process as unit tests, with the one exception that they are each compiled as their own separate crate, meaning they can access only the main crate's public interface and are run against the main crate compiled without `#[cfg(test)]`. A test harness is generated for each file in *tests/*. Test harnesses are not generated for files in subdirectories under *tests/* to allow you to have shared submodules for your tests.

NOTE *If you explicitly want a test harness for a file in a subdirectory, you can opt in to that by calling the file* main.rs.

Rust does not require that you use the default test harness. You can instead opt out of it and implement your own `main` method that represents the test runner by setting `harness = false` for a given integration test in *Cargo.toml*, as shown in Listing 6-1. The `main` method that you define will then be invoked to run the test.

```
[[test]]
name = "custom"
path = "tests/custom.rs"
harness = false
```

Listing 6-1: Opting out of the standard test harness

Without the test harness, none of the magic around #[test] happens. Instead, you're expected to write your own main function to run the testing code you want to execute. Essentially, you're writing a normal Rust binary that just happens to be run by cargo test. That binary is responsible for handling all the things that the default harness normally does (if you want to support them), such as command line flags. The harness property is set separately for each integration test, so you can have one test file that uses the standard harness and one that does not.

ARGUMENTS TO THE DEFAULT TEST HARNESS

The default test harness supports a number of command line arguments to configure how the tests are run. These aren't passed to cargo test directly but rather to the test binary that Cargo compiles and runs for you when you run cargo test. To access that set of flags, pass -- to cargo test, followed by the arguments to the test binary. For example, to see the help text for the test binary, you'd run cargo test -- --help.

A number of handy configuration options are available through these command line arguments. The --nocapture flag disables the output capturing that normally happens when you run Rust tests. This is useful if you want to observe a test's output in real time rather than all at once after the test has failed. You can use the --test-threads option to limit how many tests run concurrently, which is helpful if you have a test that hangs or segfaults and you want to figure out which one it is by running the tests sequentially. There's also a --skip option for skipping tests that match a certain pattern, --ignored to run tests that would normally be ignored (such as those that require an external program to be running), and --list to just list all the available tests.

Keep in mind that these arguments are all implemented by the default test harness, so if you disable it (with harness = false), you'll have to implement the ones you need yourself in your main function!

Integration tests without a harness are primarily useful for benchmarks, as we'll see later, but they also come in handy when you want to run tests that don't fit the standard "one function, one test" model. For example, you'll frequently see harnessless tests used with fuzzers, model checkers, and tests that require a custom global setup (like under WebAssembly or when working with custom targets).

#[cfg(test)]

When Rust builds code for testing, it sets the compiler configuration flag test, which you can then use with conditional compilation to have code that is compiled out unless it is specifically being tested. On the surface, this may seem odd: don't you want to test exactly the same code that's going into production? You do, but having code exclusively available when testing allows you to write better, more thorough tests, in a few ways.

MOCKING

When writing tests, you often want tight control over the code you're testing as well as any other types that your code may interact with. For example, if you are testing a network client, you probably do not want to run your unit tests over a real network but instead want to directly control what bytes are emitted by the "network" and when. Or, if you're testing a data structure, you want your test to use types that allow you to control what each method returns on each invocation. You may also want to gather metrics such as how often a given method was called or whether a given byte sequence was emitted.

These "fake" types and implementations are known as *mocks*, and they are a key feature of any extensive unit test suite. While you can often do the work needed to get this kind of control manually, it's nicer to have a library take care of most of the nitty-gritty details for you. This is where automated mocking comes into play. A mocking library will have facilities for generating types (including functions) with particular properties or signatures, as well as mechanisms to control and introspect those generated items during a test execution.

Mocking in Rust generally happens through generics—as long as your program, data structure, framework, or tool is generic over anything you might want to mock (or takes a trait object), you can use a mocking library to generate conforming types that will instantiate those generic parameters. You then write your unit tests by instantiating your generic constructs with the generated mock types, and you're off to the races!

In situations where generics are inconvenient or inappropriate, such as if you want to avoid making a particular aspect of your type generic to users, you can instead encapsulate the state and behavior you want to mock in a dedicated struct. You would then generate a mocked version of that struct and its methods and use conditional compilation to use either the real or mocked implementation depending on cfg(test) or a test-only feature like cfg(feature = "test_mock_foo").

At the moment, there isn't a single mocking library, or even a single mocking approach, that has emerged as the One True Answer in the Rust community. The most extensive and thorough mocking library I know of is the mockall crate, but that is still under active development, and there are many other contenders.

Test-Only APIs

First, having test-only code allows you to expose additional methods, fields, and types to your (unit) tests so the tests can check not only that the public API behaves correctly but also that the internal state is correct. For example, consider the HashMap type from hashbrown, the crate that implements the standard library HashMap. The HashMap type is really just a wrapper around a RawTable type, which is what implements most of the hash table logic. Suppose that after doing a HashMap::insert on an empty map, you want to check that a single bucket in the map is nonempty, as shown in Listing 6-2.

```
#[test]
fn insert_just_one() {
  let mut m = HashMap::new();
  m.insert(42, ());
  let full = m.table.buckets.iter().filter(Bucket::is_full).count();
  assert_eq!(full, 1);
}
```

Listing 6-2: A test that accesses inaccessible internal state and thus does not compile

This code will not compile as written, because while the test code can access the private table field of HashMap, it cannot access the also private buckets field of RawTable, as RawTable lives in a different module. We could fix this by making the buckets field visibility pub(crate), but we really don't want HashMap to be able to touch buckets in general, as it could accidentally corrupt the internal state of the RawTable. Even making buckets available as read-only could be problematic, as new code in HashMap may then start depending on the internal state of RawTable, making future modifications more difficult.

The solution is to use #[cfg(test)]. We can add a method to RawTable that allows access to buckets only while testing, as shown in Listing 6-3, and thereby avoid adding footguns for the rest of the code. The code from Listing 6-2 can then be updated to call buckets() instead of accessing the private buckets field.

```
impl RawTable {
  #[cfg(test)]
  pub(crate) fn buckets(&self) -> &[Bucket] {
    &self.buckets
  }
}
```

Listing 6-3: Using #[cfg(test)] to make internal state accessible in the testing context

Bookkeeping for Test Assertions

The second benefit of having code that exists only during testing is that you can augment the program to perform additional runtime bookkeeping that can then be inspected by tests. For example, imagine you're writing your own version of the BufWriter type from the standard library. When testing it, you want to make sure that BufWriter does not issue system calls unnecessarily. The most obvious way to do so is to have the `BufWriter` keep track of how many times it has invoked `write` on the underlying `Write`. However, in production this information isn't important, and keeping track of it introduces (marginal) performance and memory overhead. With `#[cfg(test)]`, you can have the bookkeeping happen only when testing, as shown in Listing 6-4.

```
struct BufWriter<T> {
  #[cfg(test)]
  write_through: usize,
  // other fields...
}

impl<T: Write> Write for BufWriter<T> {
  fn write(&mut self, buf: &[u8]) -> Result<usize> {
    // ...
    if self.full() {
      #[cfg(test)]
      self.write_through += 1;
      let n = self.inner.write(&self.buffer[..])?;
    // ...
  }
}
```

Listing 6-4: Using #[cfg(test)] to limit bookkeeping to the testing context

Keep in mind that test is set only for the crate that is being compiled as a test. For unit tests, this is the crate being tested, as you would expect. For integration tests, however, it is the integration test binary being compiled as a test—the crate you are testing is just compiled as a library and so will not have test set.

Doctests

Rust code snippets in documentation comments are automatically run as test cases. These are commonly referred to as *doctests*. Because doctests appear in the public documentation of your crate, and users are likely to mimic what they contain, they are run as integration tests. This means that the doctests don't have access to private fields and methods, and test is not set on the main crate's code. Each doctest is compiled as its own dedicated crate and is run in isolation, just as if the user had copy-pasted the doctest into their own program.

Behind the scenes, the compiler performs some preprocessing on doctests to make them more concise. Most importantly, it automatically adds an `fn main` around your code. This allows doctests to focus only on the important bits that the user is likely to care about, like the parts that actually use types and methods from your library, without including unnecessary boilerplate.

You can opt out of this auto-wrapping by defining your own `fn main` in the doctest. You may want to do this, for example, if you want to write an asynchronous main function using something like #[tokio::main] async fn main, or if you want to add additional modules to the doctest.

To use the ? operator in your doctest, you don't normally have to use a custom main function as rustdoc includes some heuristics to set the return type to Result<(), impl Debug> if your code looks like it makes use of ? (for example, if it ends with Ok(())). If type inference gives you a hard time about the error type for the function, you can disambiguate it by changing the last line of the doctest to be explicitly typed, like this: Ok::<(), T>(()).

Doctests have a number of additional features that come in handy as you write documentation for more complex interfaces. The first is the ability to hide individual lines. If you prefix a line of a doctest with a #, that line is included when the doctest is compiled and run, but it is not included in the code snippet generated in the documentation. This lets you easily hide details that are not important to the current example, such as implementing traits for dummy types or generating values. It is also useful if you wish to present a sequence of examples without showing the same leading code each time. Listing 6-5 gives an example of what a doctest with hidden lines might look like.

```
/// Completely frobnifies a number through I/O.
///
/// In this first example we hide the value generation.
/// ```
/// # let unfrobnified_number = 0;
/// # let already_frobnified = 1;
/// assert!(frobnify(unfrobnified_number).is_ok());
/// assert!(frobnify(already_frobnified).is_err());
/// ```
///
/// Here's an example that uses ? on multiple types
/// and thus needs to declare the concrete error type,
/// but we don't want to distract the user with that.
/// We also hide the use that brings the function into scope.
/// ```
/// # use mylib::frobnify;
/// frobnify("0".parse()?)?;
/// # Ok::<(), anyhow::Error>(())
/// ```
///
/// You could even replace an entire block of code completely,
/// though use this _very_ sparingly:
/// ```
/// # /*
/// let i = ...;
/// # */
/// # let i = 42;
/// frobnify(i)?;
/// ```
fn frobnify(i: usize) -> std::io::Result<()> {
```

Listing 6-5: Hiding lines in a doctest with #

Use this feature with care; it can be frustrating to users if they copy-paste an example and then it doesn't work because of required steps that you've hidden.

Much like #[test] functions, doctests also support attributes that modify how the doctest is run. These attributes go immediately after the triple-backtick used to denote a code block, and multiple attributes can be separated by commas.

Like with test functions, you can specify the should_panic attribute to indicate that the code in a particular doctest should panic when run, or ignore to check the code segment only if cargo test is run with the --ignored flag. You can also use the no_run attribute to indicate that a given doctest should compile but should not be run.

The attribute compile_fail tells rustdoc that the code in the documentation example should not compile. This indicates to the user that a particular use is not possible and serves as a useful test to remind you to update the documentation should the relevant aspect of your library change. You can also use this attribute to check that certain static properties hold for your types. Listing 6-6 shows an example of how you can use compile_fail to check that a given type does not implement Send, which may be necessary to uphold safety guarantees in unsafe code.

```
```compile_fail
# struct MyNonSendType(std::rc::Rc<()>);
fn is_send<T: Send>() {}
is_send::<MyNonSendType>();
```
```

Listing 6-6: Testing that code fails to compile with compile_fail

compile_fail is a fairly crude tool in that it gives no indication of *why* the code does not compile. For example, if code doesn't compile because of a missing semicolon, a compile_fail test will appear to have been successful. For that reason, you'll usually want to add the attribute only after you have made sure that the test indeed fails to compile with the expected error. If you need more fine-grained tests for compilation errors, such as when developing macros, take a look at the trybuild crate.

Additional Testing Tools

There's a lot more to testing than just running test functions and seeing that they produce the expected result. A thorough survey of testing techniques, methodologies, and tools is outside the scope of this book, but there are some key Rust-specific pieces that you should know about as you expand your testing repertoire.

Linting

You may not consider a linter's checks to be tests, but in Rust they often can be. The Rust linter *clippy* categorizes a number of its lints as *correctness lints.* These lints catch code patterns that compile but are almost certainly bugs. Some examples are `a = b; b = a`, which fails to swap a and b; `std::mem::forget(t)`, where t is a reference; and `for x in y.next()`, which will iterate only over the first element in y. If you are not running clippy as part of your CI pipeline already, you probably should be.

Clippy comes with a number of other lints that, while usually helpful, may be more opinionated than you'd prefer. For example, the `type_complexity` lint, which is on by default, issues a warning if you use a particularly involved type in your program, like `Rc<Vec<Vec<Box<(u32, u32, u32, u32)>>>>`. While that warning encourages you to write code that is easier to read, you may find it too pedantic to be broadly useful. If some part of your code erroneously triggers a particular lint, or you just want to allow a specific instance of it, you can opt out of the lint just for that piece of code with `#[allow(clippy::name_of_lint)]`.

The Rust compiler also comes with its own set of lints in the form of warnings, though these are usually more directed toward writing idiomatic code than checking for correctness. Instead, correctness lints in the compiler are simply treated as errors (take a look at `rustc -W help` for a list).

NOTE *Not all compiler warnings are enabled by default. Those disabled by default are usually still being refined, or are more about style than content. A good example of this is the "idiomatic Rust 2018 edition" lint, which you can enable with `#![warn(rust_2018 _idioms)]`. When this lint is enabled, the compiler will tell you if you're failing to take advantage of changes brought by the Rust 2018 edition. Some other lints that you may want to get into the habit of enabling when you start a new project are `missing_docs` and `missing_debug_implementations`, which warn you if you've forgotten to document any public items in your crate or add `Debug` implementations for any public types, respectively.*

Test Generation

Writing a good test suite is a lot of work. And even when you do that work, the tests you write test only the particular set of behaviors you were considering at the time you wrote them. Luckily, you can take advantage of a number of test generation techniques to develop better and more thorough tests. These generate input for you to use to check your application's correctness. Many such tools exist, each with their own strengths and weaknesses, so here I'll cover only the main strategies used by these tools: fuzzing and property testing.

Fuzzing

Entire books have been written about fuzzing, but at a high level the idea is simple: generate random inputs to your program and see if it crashes. If the program crashes, that's a bug. For example, if you're writing a URL parsing library, you can fuzz-test your program by systematically generating random strings and throwing them at the parsing function until it panics. Done naively, this would take a while to yield results: if the fuzzer starts with a, then b, then c, and so on, it will take it a long time to generate a tricky URL like http://[:]. In practice, modern fuzzers use code coverage metrics to explore different paths in your code, which lets them reach higher degrees of coverage faster than if the inputs were truly chosen at random.

Fuzzers are great at finding strange corner cases that your code doesn't handle correctly. They require little setup on your part: you just point the fuzzer at a function that takes a "fuzzable" input, and off it goes. For example, Listing 6-7 shows an example of how you might fuzz-test a URL parser.

```
libfuzzer_sys::fuzz_target!(|data: &[u8]| {
    if let Ok(s) = std::str::from_utf8(data) {
        let _ = url::Url::parse(s);
    }
});
```

Listing 6-7: Fuzzing a URL parser with libfuzzer

The fuzzer will generate semi-random inputs to the closure, and any that form valid UTF-8 strings will be passed to the parser. Notice that the code here doesn't check whether the parsing succeeds or fails—instead, it's looking for cases where the parser panics or otherwise crashes due to internal invariants that are violated.

The fuzzer keeps running until you terminate it, so most fuzzing tools come with a built-in mechanism to stop after a certain number of test cases have been explored. If your input isn't a trivially fuzzable type—something like a hash table—you can usually use a crate like arbitrary to turn the byte string that the fuzzer generates into a more complex Rust type. It feels like magic, but under the hood it's actually implemented in a very straightforward fashion. The crate defines an Arbitrary trait with a single method, arbitrary, that constructs the implementing type from a source of random bytes. Primitive types like u32 or bool read the necessary number of bytes from that input to construct a valid instance of themselves, whereas more complex types like HashMap or BTreeSet produce one number from the input to dictate their length and then call Arbitrary that number of times on their inner types. There's even an attribute, #[derive(Arbitrary)], that implements Arbitrary by just calling arbitrary on each contained type! To explore fuzzing further, I recommend starting with cargo-fuzz.

Property-Based Testing

Sometimes you want to check not only that your program doesn't crash but also that it does what it's expected to do. It's great that your add function didn't panic, but if it tells you that the result of add(1, 4) is 68, it's probably still wrong. This is where *property-based testing* comes into play; you describe a number of properties your code should uphold, and then the property testing framework generates inputs and checks that those properties indeed hold.

A common way to use property-based testing is to first write a simple but naive version of the code you want to test that you are confident is correct. Then, for a given input, you give that input to both the code you want to test and the simplified but naive version. If the result or output of the two implementations is the same, your code is good—that is the correctness property you're looking for—but if it's not, you've likely found a bug. You can also use property-based testing to check for properties not directly related to correctness, such as whether operations take strictly less time for one implementation than another. The common principle is that you want any difference in outcome between the real and test versions to be informative and actionable so that every failure allows you to make improvements. The naive implementation might be one from the standard library that you're trying to replace or augment (like std::collections::VecDeque), or it might be a simpler version of an algorithm that you're trying optimize (like naive versus optimized matrix multiplication).

If this approach of generating inputs until some condition is met sounds a lot like fuzzing, that's because it is—smarter people than I have argued that fuzzing is "just" property-based testing where the property you're testing for is "it doesn't crash."

One downside of property-based testing is that it relies more heavily on the provided descriptions of the inputs. Whereas fuzzing will keep trying all possible inputs, property testing tends to be guided by developer annotations like "a number between 0 and 64" or "a string that contains three commas." This allows property testing to more quickly reach cases that fuzzers may take a long time to encounter randomly, but it does require manual work and may miss important but niche buggy inputs. As fuzzers and property testers grow closer, however, fuzzers are starting to gain this kind of constraint-based searching capability as well.

If you're curious about property-based test generation, I recommend starting with the proptest crate.

TESTING SEQUENCES OF OPERATIONS

Since fuzzers and property testers allow you to generate arbitrary Rust types, you aren't limited to testing a single function call in your crate. For example, say you want to test that some type Foo behaves correctly if you perform a particular sequence of operations on it. You could define an enum Operation that lists operations, and make your test function take a Vec<Operation>. Then you could instantiate a Foo and perform each operation on that Foo, one after the other. Most testers have support for minimizing inputs, so they will even search for the smallest sequence of operations that still violates a property if a property-violating input is found!

Test Augmentation

Let's say you have a magnificent test suite all set up, and your code passes all the tests. It's glorious. But then, one day, one of the normally reliable tests inexplicably fails or crashes with a segmentation fault. There are two common reasons for these kinds of nondeterministic test failures: race conditions, where your test might fail only if two operations occur on different threads in a particular order, and undefined behavior in unsafe code, such as if some unsafe code reads a particular value out of uninitialized memory.

Catching these kinds of bugs with normal tests can be difficult—often you don't have sufficient low-level control over thread scheduling, memory layout and content, or other random-ish system factors to write a reliable test. You could run each test many times in a loop, but even that may not catch the error if the bad case is sufficiently rare or unlikely. Luckily, there are tools that can help augment your tests to make catching these kinds of bugs much easier.

The first of these is the amazing tool *Miri*, an interpreter for Rust's *mid-level intermediate representation (MIR)*. MIR is an internal, simplified representation of Rust that helps the compiler find optimizations and check properties without having to consider all of the syntax sugar of Rust itself. Running your tests through Miri is as simple as running `cargo miri test`. Miri *interprets* your code rather than compiling and running it like a normal binary, which makes the tests run a decent amount slower. But in return, Miri can keep track of the entire program state as each line of your code executes. This allows Miri to detect and report if your program ever exhibits certain types of undefined behavior, such as uninitialized memory reads, uses of values after they've been dropped, or out-of-bounds pointer accesses. Rather than having these operations yield strange program behaviors that may only sometimes result in observable test failures (like crashes), Miri detects them when they happen and tells you immediately.

For example, consider the very unsound code in Listing 6-8, which creates two exclusive references to a value.

```
let mut x = 42;
let x: *mut i32 = &mut x;
let (x1, x2) = unsafe { (&mut *x, &mut *x) };
println!("{} {}", x1, x2);
```

Listing 6-8: Wildly unsafe code that Miri detects is incorrect

At the time of writing, if you run this code through Miri, you get an error that points out exactly what's wrong:

```
error: Undefined Behavior: trying to reborrow for Unique at alloc1383, but
parent tag <2772> does not have an appropriate item in the borrow stack
 --> src/main.rs:4:6
  |
4 | let (x1, x2) = unsafe { (&mut *x, &mut *x) };
  |      ^^ trying to reborrow for Unique at alloc1383, but parent tag <2772>
does not have an appropriate item in the borrow stack
```

Miri is still under development, and its error messages aren't always the easiest to understand. This is a problem that's being actively worked on, so by the time you read this, the error output may have already gotten much better!

Another tool worth looking at is *Loom*, a clever library that tries to ensure your tests are run with every relevant interleaving of concurrent operations. At a high level, Loom keeps track of all cross-thread synchronization points and runs your tests over and over, adjusting the order in which threads proceed from those synchronization points each time. So, if thread A and thread B both take the same Mutex, Loom will ensure that the test runs once with A taking it first and once with B taking it first. Loom also keeps track of atomic accesses, memory orderings, and accesses to UnsafeCell (which we'll discuss in Chapter 9) and checks that threads do not access them inappropriately. If a test fails, Loom can give you an exact rundown of which threads executed in what order so you can determine how the crash happened.

Performance Testing

Writing performance tests is difficult because it is often hard to accurately model a workload that reflects real-world usage of your crate. But having such tests is important; if your code suddenly runs 100 times slower, that really should be considered a bug, yet without a performance test you may not spot the regression. If your code runs 100 times *faster,* that might also indicate that something is off. Both of these are good reasons to have automated performance tests as part of your CI—if performance changes drastically in either direction, you should know about it.

Unlike with functional testing, performance tests do not have a common, well-defined output. A functional test will either succeed or fail, whereas a performance test may give you a throughput number, a latency profile, a number of processed samples, or any other metric that might be relevant to the application in question. Also, a performance test may require running a function in a loop a few hundred thousand times, or it might take hours running across a distributed network of multicore boxes. For that reason, it is difficult to speak about how to write performance tests in a general sense. Instead, in this section, we'll look at some of the issues you may encounter when writing performance tests in Rust and how to mitigate them. Three particularly common pitfalls that are often overlooked are performance variance, compiler optimizations, and I/O overhead. Let's explore each of these in turn.

Performance Variance

Performance can vary for a huge variety of reasons, and many factors affect how fast a particular sequence of machine instructions run. Some are obvious, like the CPU and memory clock speed, or how loaded the machine otherwise is, but many are much more subtle. For example, your kernel version may change paging performance, the length of your username might change the layout of memory, and the temperature in the room might cause the CPU to clock down. Ultimately, it is highly unlikely that if you run a benchmark twice, you'll get the same result. In fact, you may observe significant variance, even if you are using the same hardware. Or, viewed from another perspective, your code may have gotten slower or faster, but the effect may be invisible due to differences in the benchmarking environment.

There are no perfect ways to eliminate all variance in your performance results, unless you happen to be able to run benchmarks repeatedly on a highly diverse fleet of machines. Even so, it's important to try to handle this measurement variance as best we can to extract a signal from the noisy measurements benchmarks give us. In practice, our best friend in combating variance is to run each benchmark many times and then look at the *distribution* of measurements rather than just a single one. Rust has tools that can help with this. For example, rather than ask "How long did this function take to run on average?" crates like hdrhistogram enable us to look at statistics like "What range of runtime covers 95% of the samples we observed?" To be even more rigorous, we can use techniques like null hypothesis testing from statistics to build some confidence that a measured difference indeed corresponds to a true change and is not just noise.

A lecture on statistical hypothesis testing is beyond the scope of this book, but luckily much of this work has already been done by others. The criterion crate, for instance, does all of this and more for you. All you have to do is give it a function that it can call to run one iteration of your benchmark, and it will run it the appropriate number of times to be fairly sure that the result is reliable. It then produces a benchmark report, which includes a summary of the results, analysis of outliers, and even graphical representations of trends over time. Of course, it can't eliminate the effects of just testing on a particular configuration of hardware, but it at least categorizes the noise that is measurable across executions.

Compiler Optimizations

Compilers these days are really clever. They eliminate dead code, compute complex expressions at compile time, unroll loops, and perform other dark magic to squeeze every drop of performance out of our code. Normally this is great, but when we're trying to measure how fast a particular piece of code is, the compiler's smartness can give us invalid results. For example, take the code to benchmark Vec::push in Listing 6-9.

```
let mut vs = Vec::with_capacity(4);
let start = std::time::Instant::now();
for i in 0..4 {
  vs.push(i);
}
println!("took {:?}", start.elapsed());
```

Listing 6-9: A suspiciously fast performance benchmark

If you were to look at the assembly output of this code compiled in release mode using something like the excellent *godbolt.org* or cargo-asm, you'd immediately notice that something was wrong: the calls to Vec::with _capacity and Vec::push, and indeed the whole for loop, are nowhere to be seen. They have been optimized out completely. The compiler realized that nothing in the code actually required the vector operations to be performed and eliminated them as dead code. Of course, the compiler is completely within its rights to do so, but for benchmarking purposes, this is not particularly helpful.

To avoid these kinds of optimizations for benchmarking, the standard library provides std::hint::black_box. This function has been the topic of much debate and confusion and is still pending stabilization at the time of writing, but is so useful it's worth discussing here nonetheless. At its core, it's simply an identity function (one that takes x and returns x) that tells the compiler to assume that the argument to the function is used in arbitrary (legal) ways. It does not prevent the compiler from applying optimizations to the input argument, nor does it prevent the compiler from optimizing how the return value is used. Instead, it encourages the compiler to actually compute the argument to the function (under the assumption that it will be used) and to store that result somewhere accessible to the CPU such that black_box could be called with the computed value. The compiler is free to, say, compute the input argument at compile time, but it should still inject the result into the program.

This function is all we need for many, though admittedly not all, of our benchmarking needs. For example, we can annotate Listing 6-9 so that the vector accesses are no longer optimized out, as shown in Listing 6-10.

```
let mut vs = Vec::with_capacity(4);
let start = std::time::Instant::now();
for i in 0..4 {
  black_box(vs.as_ptr());
  vs.push(i);
  black_box(vs.as_ptr());
}
println!("took {:?}", start.elapsed());
```

Listing 6-10: A corrected version of Listing 6-9

We've told the compiler to assume that vs is used in arbitrary ways on each iteration of the loop, both before and after the calls to push. This forces the compiler to perform each push in order, without merging or otherwise optimizing consecutive calls, since it has to assume that "arbitrary stuff that cannot be optimized out" (that's the black_box part) may happen to vs between each call.

Note that we used vs.as_ptr() and not, say, &vs. That's because of the caveat that the compiler should assume black_box can perform any *legal* operation on its argument. It is not legal to mutate the Vec through a shared reference, so if we used black_box(&vs), the compiler might notice that vs will not change between iterations of the loop and implement optimizations based on that observation!

I/O Overhead Measurement

When writing benchmarks, it's easy to accidentally measure the wrong thing. For example, we often want to get information in real time about how far along the benchmark is. To do that, we might write code like that in Listing 6-11, intended to measure how fast my_function runs:

```
let start = std::time::Instant::now();
for i in 0..1_000_000 {
  println!("iteration {}", i);
  my_function();
}
println!("took {:?}", start.elapsed());
```

Listing 6-11: What are we really benchmarking here?

This may look like it achieves the goal, but in reality, it does not actually measure how fast my_function is. Instead, this loop is most likely to tell us how long it takes to print a million numbers. The println! in the body of the loop does a lot of work behind the scenes: it turns a binary integer into decimal digits for printing, locks standard output, writes out a sequence of UTF-8 code points using at least one system call, and then releases the standard output lock. Not only that, but the system call might block if your terminal is slow to print out the input it receives. That's a lot of cycles! And the time it takes to call my_function might pale in comparison.

A similar thing happens when your benchmark uses random numbers. If you run my_function(rand::random()) in a loop, you may well be mostly measuring the time it takes to generate a million random numbers. The story is the same for getting the current time, reading a configuration file, or starting a new thread—these things all take a long time, relatively speaking, and may end up overshadowing the time you actually wanted to measure.

Luckily, this particular issue is often easy to work around once you are aware of it. Make sure that the body of your benchmarking loop contains almost nothing but the particular code you want to measure. All other code should run either before the benchmark begins or outside of the measured part of the benchmark. If you're using criterion, take a look at the different timing loops it provides—they're all there to cater to benchmarking cases that require different measurement strategies!

Summary

In this chapter, we explored the built-in testing capabilities that Rust offers in great detail. We also looked at a number of testing facilities and techniques that are useful when testing Rust code. This is the last chapter that focuses on higher-level aspects of intermediate Rust use in this book. Starting with the next chapter on declarative and procedural macros, we will be focusing much more on Rust code. See you on the next page!

7

MACROS

Macros are, in essence, a tool for making the compiler write code for you. You give the compiler a formula for generating code given some input parameters, and the compiler replaces every invocation of the macro with the result of running through the formula. You can think of macros as automatic code substitution where you get to define the rules for the substitution.

Rust's macros come in many different shapes and sizes to make it easy to implement many different forms of code generation. The two primary types are *declarative* macros and *procedural* macros, and we will explore both of them in this chapter. We'll also look at some of the ways macros can come in handy in your everyday coding and some of the pitfalls that arise with more advanced use.

Programmers coming from C-based languages may be used to the unholy land of C and C++ where you can use #define to change each true to false, or to remove all occurrences of the else keyword. If that's the case for you, you'll need to disassociate macros from a feeling of doing something "bad." Macros in Rust are far from the Wild West of C macros. They follow (mostly) well-defined rules and are fairly misuse-resistant.

Declarative Macros

Declarative macros are those defined using the macro_rules! syntax, which lets you conveniently define function-like macros without having to resort to writing a dedicated crate for the purpose (as you do with procedural macros). Once you've defined a declarative macro, you can invoke it using the name of the macro followed by an exclamation mark. I like to think of this kind of macro as a sort of compiler-assisted search and replace: it does the job for many regular, well-structured transformation tasks, and for eliminating repetitive boilerplate. In your experience with Rust up until this point, most of the macros you have recognized as macros are likely to have been declarative macros. Note, however, that not all function-like macros are declarative macros; macro_rules! itself is one example of this, and format_args! is another. The ! suffix merely indicates to the compiler that the macro invocation will be replaced with different source code at compile time.

NOTE *Since Rust's parser specifically recognizes and parses macro invocations annotated with !, you can use them only in places where the parser allows them. They work in most places you'd expect, like in expression position or in an impl block, but not everywhere. For example, you cannot (at the time of writing) invoke a function-like macro where an identifier or match arm is expected.*

It may not be immediately obvious why declarative macros are called declarative. After all, don't you "declare" everything in your program? In this context, *declarative* refers to the fact that you don't say *how* the macro's inputs should be translated into the output, just that you want the output to look like A when the input is B. You declare that it shall be so, and the compiler figures out all the parsing rewiring that has to happen to make your declaration reality. This makes declarative macros concise and expressive, though it also has a tendency to make them rather cryptic since you have a limited language with which to express your declarations.

When to Use Them

Declarative macros are primarily useful when you find yourself writing the same code over and over, and you'd like to, well, not do that. They're best suited for fairly mechanical replacements—if you're aiming to do fancy code transformations or lots of code generation, procedural macros are likely a better fit.

I most frequently use declarative macros in cases where I find myself writing repetitive and structurally similar code, such as in tests and trait implementations. For tests, I often want to run the same test multiple times but with slightly different configurations. I might have something like what is shown in Listing 7-1.

```
fn test_inner<T>(init: T, frobnify: bool) { ... }
#[test]
fn test_1u8_frobnified() {
  test_inner(1u8, true);
}
// ...
#[test]
fn test_1i128_not_frobnified() {
  test_inner(1i128, false);
}
```

Listing 7-1: Repetitive testing code

While this works, it's too verbose, too repetitive, and too prone to manual error. With macros we can do much better, as shown in Listing 7-2.

```
macro_rules! test_battery {
  ($($t:ty as $name:ident),*)) => {
    $(
      mod $name {
        #[test]
        fn frobnified() { test_inner::<$t>(1, true) }
        #[test]
        fn unfrobnified() { test_inner::<$t>(1, false) }
      }
    )*
  }
}
test_battery! {
  u8 as u8_tests,
  // ...
  i128 as i128_tests
);
```

Listing 7-2: Making a macro repeat for you

This macro expands each comma-separated directive into its own module that then contains two tests, one that calls test_inner with true, and one with false. While the macro definition isn't trivial, it makes adding more tests much easier. Each type is one line in the test_battery! invocation, and the macro will take care of generating tests for both true and false arguments. We could also have it generate tests for different values for init. We've now significantly reduced the likelihood that we'll forget to test a particular configuration!

The story for trait implementations is similar. If you define your own trait, you'll often want to implement that trait for a number of types in the standard library, even if those implementations are trivial. Let's imagine you invented the Clone trait and want to implement it for all the Copy types in the standard library. Instead of manually writing an implementation for each one, you can use a macro like the one in Listing 7-3.

```
macro_rules! clone_from_copy {
  ($($t:ty),*) => {
    $(impl Clone for $t {
      fn clone(&self) -> Self { *self }
    })*
  }
}
clone_from_copy![bool, f32, f64, u8, i8, /* ... */];
```

Listing 7-3: Using a macro to implement a trait for many similar types in one fell swoop

Here, we generate an implementation of Clone for each provided type whose body just uses * to copy out of &self. You may wonder why we don't add a blanket implementation of Clone for T where T: Copy. We could do that, but a big reason not to is that it would force types in other crates to also use that same implementation of Clone for their own types that happen to be Copy. An experimental compiler feature called *specialization* could offer a workaround, but at the time of writing the stabilization of that feature is still some way off. So, for the time being, we're better off enumerating the types specifically. This pattern also extends beyond simple forwarding implementations: for example, you could easily alter the code in Listing 7-3 to implement an AddOne trait to all integer types!

NOTE *If you ever find yourself wondering if you should use generics or a declarative macro, you should use generics. Generics are generally more ergonomic than macros and integrate much better with other constructs in the language. Consider this rule of thumb: if your code changes based on type, use generics; otherwise, use macros.*

How They Work

Every programming language has a *grammar* that dictates how the individual characters that make up the source code can be turned into *tokens*. Tokens are the lowest-level building blocks of a language, such as numbers, punctuation characters, string and character literals, and identifiers; at this level, there's no distinction between language keywords and variable names. For example, the text (value + 4) would be represented by the five-token sequence (, value, +, 4,) in Rust-like grammar. The process of turning text into tokens also provides a layer of abstraction between the rest of the compiler and the gnarly low-level details of parsing text. For example, in the token representation, there is no notion of whitespace, and /*"foo"*/ and "/*foo*/" have distinct representations (the former is no token, and the latter is a string literal token with the content /*foo*/).

Once the source code has been turned into a sequence of tokens, the compiler walks that sequence and assigns syntactic meaning to the tokens. For example, ()-delimited tokens make up a group, ! tokens denote macro invocations, and so on. This is the process of *parsing*, which ultimately produces an abstract syntax tree (AST) that describes the structure represented by the source code. As an example, consider the expression let x = || 4, which consists of the sequence of tokens let (keyword), x (identifier), = (punctuation), two instances of | (punctuation), and 4 (literal). When the compiler turns that into a syntax tree, it represents it as a *statement* whose *pattern* is the *identifier* x and whose right-hand *expression* is a *closure* that has an empty *argument list* and a *literal expression* of the *integer literal* 4 as its body. Notice how the syntax tree representation is much richer than the token sequence, since it assigns syntactic meaning to the token combinations following the language's grammar.

Rust macros dictate the syntax tree that a given sequence of tokens gets turned into—when the compiler encounters a macro invocation during parsing, it has to evaluate the macro to determine the replacement tokens, which will ultimately become the syntax tree for the macro invocation. At this point, however, the compiler is still parsing the tokens and might not be in a position to evaluate a macro yet, since all it has done is parse the tokens of the macro definition. Instead, then, the compiler defers the parsing of anything contained within the delimiters of a macro invocation and remembers the input token sequence. When the compiler is ready to evaluate the indicated macro, it evaluates the macro over the token sequence, parses the tokens it yields, and substitutes the resulting syntax tree into the tree where the macro invocation was.

Technically, the compiler does do a little bit of parsing for the input to a macro. Specifically, it parses out basic things like string literals and delimited groups and so produces a sequence of token *trees* rather than just tokens. For example, the code x - (a.b + 4) parses as a sequence of three token trees. The first token tree is a single token that is the identifier x, the second is a single token that is the punctuation character -, and the third is a group (using parentheses as the delimiter), which itself consists of a sequence of five token trees: a (an identifier), . (punctuation), b (another identifier), + (another punctuation token), and 4 (a literal). This means that the input to a macro does not necessarily have to be valid Rust, but it must consist of code that the Rust compiler can parse. For example, you couldn't write for <- x in Rust outside of a macro invocation, but inside of a macro invocation you can, as long as the macro produces valid syntax. On the other hand, you cannot pass for { to a macro because it doesn't have a closing brace.

Declarative macros always generate valid Rust as output. You cannot have a macro generate, say, the first half of a function invocation or an if without the block that follows it. A declarative macro must generate an expression (basically anything that you can assign to a variable), a statement such as let x = 1;, an item like a trait definition or impl block, a type, or a match pattern. This makes Rust macros resistant to misuse: you simply cannot write a declarative macro that generates invalid Rust code, because the macro definition itself would not compile!

That's really all there is to declarative macros at a high level—when the compiler encounters a macro invocation, it passes the tokens contained within the invocation delimiters to the macro, parses the resulting token stream, and replaces the macro invocation with the resulting AST.

How to Write Declarative Macros

An exhaustive explanation of all the syntax that declarative macros support is outside the scope of this book. However, we'll cover the basics as there are some oddities worth pointing out.

Declarative macros consist of two main parts: *matchers* and *transcribers*. A given macro can have many matchers, and each matcher has an associated transcriber. When the compiler finds a macro invocation, it walks the macro's matchers from first to last, and when it finds a matcher that matches the tokens in the invocation, it substitutes the invocation by walking the tokens of the corresponding transcriber. Listing 7-4 shows how the different parts of a declarative macro rule fit together.

```
macro_rules! /* macro name */ {
    (/* 1st matcher */) => { /* 1st transcriber */ };
    (/* 2nd matcher */) => { /* 2nd transcriber */ };
}
```

Listing 7-4: Declarative macro definition components

Matchers

You can think of a macro matcher as a token tree that the compiler tries to twist and bend in predefined ways to match the input token tree it was given at the invocation site. As an example, consider a macro with the matcher `$a:ident + $b:expr`. That matcher will match any identifier (`:ident`) followed by a plus sign followed by any Rust expression (`:expr`). If the macro is invoked with `x + 3 * 5`, the compiler notices that the matcher matches if it sets `$a = x` and `$b = 3 * 5`. Even though `*` never appears in the matcher, the compiler realizes that `3 * 5` is a valid expression and that it can therefore be matched with `$b:expr`, which accepts anything that is an expression (the `:expr` part).

Matchers can get pretty hairy, but they have huge expressive power, much like regular expressions. For a not-too-hairy example, this matcher accepts a sequence (`$()`) of one or more (`+`) comma-separated (`),`) key/value pairs given in key => value format:

```
$($key:expr => $value:expr),+
```

And, crucially, code that invokes a macro with this matcher can give an arbitrarily complex expression for the key or value—the magic of matchers will make sure that the key and value expressions are partitioned appropriately.

Macro rules support a wide variety of *fragment types*; you've already seen `:ident` for identifiers and `:expr` for expressions, but there is also `:ty` for types and even `:tt` for any single token tree! You can find a full list of the fragment types in Chapter 3 of the Rust language reference (*https://doc.rust-lang.org/reference/macros-by-example.html*). These, plus the mechanism for matching a pattern repeatedly (`$()`), enable you to match most straightforward code patterns. If, however, you find that it is difficult to express the pattern you want with a matcher, you may want to try a procedural macro instead, where you don't need to follow the strict syntax that `macro_rules!` requires. We'll look at these in more detail later in the chapter.

Transcribers

Once the compiler has matched a declarative macro matcher, it generates code using the matcher's associated transcriber. The variables defined by a macro matcher are called *metavariables,* and the compiler substitutes any occurrence of each metavariable in the transcriber (like $key in the example in the previous section) with the input that matches that part of the matcher. If you have repetition in the matcher (like $(),+ in that same example), you can use the same syntax in the transcriber and it will be repeated once for each match in the input, with each expansion holding the appropriate substitution for each metavariable for that iteration. For example, for the $key and $value matcher, we could write the following transcriber to generate an insert call into some map for each $key/$value pair that was matched:

```
$(map.insert($key, $value);)+
```

Notice that here we want a semicolon for each repetition, not just to delimit the repetition, so we place the semicolon inside the repetition parentheses.

NOTE *You must use a metavariable in each repetition in the transcriber so that the compiler knows which repetition in the matcher to use (in case there is more than one).*

Hygiene

You may have heard that Rust macros are *hygienic,* and perhaps that being hygienic makes them safer or nicer to work with, without necessarily understanding what that means. When we say Rust macros are hygienic, we mean that a declarative macro (generally) cannot affect variables that aren't explicitly passed to it. A trivial example is that if you declare a variable with the name foo, and then call a macro that also defines a variable named foo, the macro's foo is by default not visible at the call site (the place where the macro is called from). Similarly, macros cannot access variables defined at the call site (even self) unless they are explicitly passed in.

You can, most of the time, think of macro identifiers as existing in their own namespace that is separate from that of the code they expand into. For an example, take a look at the code in Listing 7-5, which has a macro that tries (and fails) to shadow a variable at the call site.

```
macro_rules! let_foo {
  ($x:expr) => {
    let foo = $x;
  }
}
let foo = 1;
// expands to let foo = 2;
let_foo!(2);
assert_eq!(foo, 1);
```

Listing 7-5: Macros exist in their own little universes. Mostly.

After the compiler expands let_foo!(2), the assert looks like it should fail. However, the foo from the original code and the one generated by the macro exist in different universes and have no relationship to one another beyond that they happen to share a human-readable name. In fact, the compiler will complain that the let foo in the macro is an unused variable. This hygiene is very helpful in making macros easier to debug—you don't have to worry about accidentally shadowing or overwriting variables in the macro caller just because you happened to choose the same variable names!

This hygienic separation does not apply beyond variable identifiers, however. Declarative macros do share a namespace for types, modules, and functions with the call site. This means your macro can define new functions that can be called in the invoking scope, add new implementations to a type defined elsewhere (and not passed in), introduce a new module that can then be accessed where the macro was invoked, and so on. This is by design—if macros could not affect the broader code like this, it would be much more cumbersome to use them to generate types, trait implementations, and functions, which is where they come in most handy.

The lack of hygiene for types in macros is particularly important when writing a macro you want to export from your crate. For the macro to truly be reusable, you cannot assume anything about what types will be in scope at the caller. Maybe the code that calls your macro has a mod std {} defined or has imported its own Result type. To be on the safe side, make sure you use fully specified types like ::core::option::Option or ::alloc::boxed::Box. If you specifically need to refer to something in the crate that defines the macro, use the special metavariable $crate.

NOTE   *Avoid using ::std paths if you can so that the macro will continue to work in no_std crates.*

You can choose to share identifiers between a macro and its caller if you want the macro to affect a specific variable in the caller's scope. The key is to remember where the identifier originated, because that's the namespace the identifier will be tied to. If you put let foo = 1; in a macro, the identifier foo originates in the macro and will never be available to the identifier namespace at the caller. If, on the other hand, the macro takes $foo:ident as an argument and then writes let $foo = 1;, when the caller invokes the macro with !(foo), the identifier will have originated in the caller and will therefore refer to foo in the caller's scope.

The identifier does not have to be quite so explicitly passed, either; any identifier that appears in code that originates outside the macro will refer to the identifier in the caller's scope. In the example in Listing 7-6, the variable identifier appears in an :expr but nonetheless has access to the variable in the caller's scope.

```
macro_rules! please_set {
    ($i:ident, $x:expr) => {
        $i = $x;
    }
```

```
}
let mut x = 1;
please_set!(x, x + 1);
assert_eq!(x, 2);
```

Listing 7-6: Giving macros access to identifiers at the call site

We could have used = $i + 1 in the macro instead, but we could not have used = x + 1 as the name x is not available in the macro's definition scope.

One last note on declarative macros and scoping: unlike pretty much everything else in Rust, declarative macros exist in the source code only after they are declared. If you try to use a macro that you define further down in the file, this will not work! This applies globally to your project; if you declare a macro in one module and want to use it in another, the module you declare the macro in must appear earlier in the crate, not later. If foo and bar are modules at the root of a crate, and foo declares a macro that bar wants to use, then mod foo must appear before mod bar in *lib.rs*!

NOTE *There is one exception to this odd scoping of macros (formally called textual scoping), and that is if you mark the macro with #[macro_export]. That annotation effectively hoists the macro to the root of the crate and marks it as pub so that it can then be used anywhere in your crate or by your crate's dependents.*

Procedural Macros

You can think of a procedural macro as a combination of a parser and code generation, where you write the glue code in between. At a high level, with procedural macros, the compiler gathers the sequence of input tokens to the macro and runs your program to figure out what tokens to replace them with.

Procedural macros are so called because you define *how* to generate code given some input tokens rather than just writing what code gets generated. There are very few smarts involved on the compiler's side—as far as it is aware, the procedural macro is more or less a source code preprocessor that may replace code arbitrarily. The requirement that your input can be parsed as a stream of Rust tokens still holds, but that's about it!

Types of Procedural Macros

Procedural macros come in three different flavors, each specialized to a particular common use case:

- Function-like macros, like the ones that macro_rules! generates
- Attribute macros, like #[test]
- Derive macros, like #[derive(Serialize)]

All three types use the same underlying mechanism: the compiler provides your macro with a sequence of tokens, and it expects you to produce a sequence of tokens in return that are (probably) related to the input tree. However, they differ in how the macro is invoked and how its output is handled. We'll cover each one briefly.

Function-Like Macros

The function-like macro is the simplest form of procedural macro. Like a declarative macro, it simply replaces the macro code at the call site with the code that the procedural macro returns. However, unlike with declarative macros, all the guard rails are off: these macros (like all procedural macros) are not required to be hygienic and will not protect you from interacting with identifiers in the surrounding code at the call site. Instead, your macros are expected to explicitly call out which identifiers should overlap with the surrounding code (using Span::call_site) and which should be treated as private to the macro (using Span::mixed_site, which we'll discuss later).

Attribute Macros

The attribute macro also replaces the item that the attribute is assigned to wholesale, but this one takes two inputs: the token tree that appears in the attribute (minus the attribute's name) and the token tree of the entire item it is attached to, including any other attributes that item may have. Attribute macros allow you to easily write a procedural macro that transforms an item, such as by adding a prelude or epilogue to a function definition (like #[test] does) or by modifying the fields of a struct.

Derive Macros

The derive macro is slightly different from the other two in that it adds to, rather than replaces, the target of the macro. Even though this limitation may seem severe, derive macros were one of the original motivating factors behind the creation of procedural macros. Specifically, the serde crate needed derive macros to be able to implement its now-well-known #[derive(Serialize, Deserialize)] magic.

Derive macros are arguably the simplest of the procedural macros, since they have such a rigid form: you can append items only after the annotated item; you can't replace the annotated item, and you cannot have the derivation take arguments. Derive macros do allow you to define *helper attributes*—attributes that can be placed inside the annotated type to give clues to the derive macro (like #[serde(skip)])—but these function mostly like markers and are not independent macros.

The Cost of Procedural Macros

Before we talk about when each of the different procedural macro types is appropriate, it's worth discussing why you may want to think twice before you reach for a procedural macro—namely, increased compile time.

Procedural macros can significantly increase compile times for two main reasons. The first is that they tend to bring with them some pretty heavy dependencies. For example, the syn crate, which provides a parser for Rust token streams that makes the experience of writing procedural macros much easier, can take tens of seconds to compile with all features enabled. You can (and should) mitigate this by disabling features you do not need and compiling your procedural macros in debug mode rather than release mode. Code often compiles several times faster in debug mode, and for most procedural macros, you won't even notice the difference in execution time.

The second reason why procedural macros increase compile time is that they make it easy for you to generate a lot of code without realizing it. While the macro saves you from having to actually type the generated code, it does not save the compiler from having to parse, compile, and optimize it. As you use more procedural macros, that generated boilerplate adds up, and it can bloat your compile times.

That said, the actual execution time of procedural macros is rarely a factor in overall compile time. While the compiler has to wait for the procedural macro to do its thing before it can continue, in practice, most procedural macros don't do any heavy computation. That said, if your procedural macro is particularly involved, you may end up with your compiles spending a significant chunk of execution time on your procedural macro code, which is worth keeping an eye out for!

So You Think You Want a Macro

Let's now look at some good uses for each type of procedural macro. We'll start with the easy one: derive macros.

When to Use Derive Macros

Derive macros are used for one thing, and one thing only: to automate the implementation of a trait where automation is possible. Not all traits have obvious automated implementations, but many do. In practice, you should consider adding a derive macro for a trait only if the trait is implemented often and if its implementation for any given type is fairly obvious. The first of these conditions may seem like common sense; if your trait is going to be implemented only once or twice, it's probably not worth writing and maintaining a convoluted derive macro for it.

The second condition may seem stranger, however: what does it mean for the implementation to be "obvious"? Consider a trait like Debug. If you were told what Debug does and were shown a type, you would probably expect an implementation of Debug to output the name of each field alongside the debug representation of its value. And that's what derive(Debug) does. What about Clone? You'd probably expect it to just clone every field— and again, that's what derive(Clone) does. With derive(serde::Serialize), we expect it to serialize every field and its value, and it does just that. In general, you want the derivation of a trait to match the developer's intuition for what it probably does. If there is no obvious derivation for a trait, or worse yet, if your derivation does not match the obvious implementation, then you're probably better off not giving it a derive macro.

When to Use Function-Like Macros

Function-like macros are harder to give a general rule of thumb for. You might say you should use function-like macros when you want a function-like macro but can't express it with `macro_rules!`, but that's a fairly subjective guideline. You can do a lot with declarative macros if you really put your mind to it, after all!

There are two particularly good reasons to reach for a function-like macro:

- If you already have a declarative macro, and its definition is becoming so hairy that the macro is hard to maintain.
- If you have a pure function that you need to be able to execute at compile time but cannot express it with `const fn`. An example of this is the `phf` crate, which generates a hash map or set using a perfect hash function when given a set of keys provided at compile time. Another is `hex-literal`, which takes a string of hexadecimal characters and replaces it with the corresponding bytes. In general, anything that does not merely transform the input at compile time but actually computes over it is likely to be a good candidate.

I do not recommend reaching for a function-like macro just so that you can break hygiene within your macro. Hygiene for function-like macros is a feature that avoids many debugging headaches, and you should think very carefully before you intentionally break it.

When to Use Attribute Macros

That leaves us with attribute macros. Though these are arguably the most general of procedural macros, they are also the hardest to know when to use. Over the years and time and time again, I have seen four ways in which attribute macros add tremendous value.

Test generation

It is very common to want to run the same test under multiple different configurations, or many similar tests with the same bootstrapping code. While a declarative macro may let you express this, your code is often easier to read and maintain if you have an attribute like #[foo_test] that introduces a setup prelude and postscript in each annotated test, or a repeatable attribute like #[test_case(1)] #[test_case(2)] to mark that a given test should be repeated multiple times, once with each input.

Framework annotations

Libraries like rocket use attribute macros to augment functions and types with additional information that the framework then uses without the user having to do a lot of manual configuration. It's so much more convenient to be able to write #[get("/<name>")] fn hello(name: String) than to have to set up a configuration struct with function pointers and the like. Essentially, the attributes make up a miniature domain-specific language (DSL) that hides a lot of boilerplate that'd otherwise be necessary. Similarly, the asynchronous I/O framework tokio lets you use #[tokio::main] async fn main() to automatically set up a runtime and run your asynchronous code, thereby saving you from writing the same runtime setup in every asynchronous application's main function.

Transparent middleware

Some libraries want to inject themselves into your application in unobtrusive ways to provide added value that does not change the application's functionality. For example, tracing and logging libraries like tracing and metric collection libraries like metered allow you to transparently instrument a function by adding an attribute to it, and then every call to that function will run some additional code dictated by the library.

Type transformers

Sometimes you want to go beyond merely deriving traits for a type and actually change the type's definition in some fundamental way. In these cases, attribute macros are the way to go. The pin_project crate is a great example of this: its primary purpose is not to implement a particular trait but rather to ensure that all pinned access to fields of a given type happens according to the strict rules that are set forth by Rust's Pin type and the Unpin trait (we'll talk more about those types in Chapter 8). It does this by generating additional helper types, adding methods to the annotated type, and introducing static safety checks to ensure that users don't accidentally shoot themselves in the foot. While pin_project could have been implemented with a procedural derive macro, that derived trait implementation would likely not have been obvious, which violates one of our rules for when to use procedural macros.

How Do They Work?

At the heart of all procedural macros is the TokenStream type, which can be iterated over to get the individual TokenTree items that make up that token stream. A TokenTree is either a single token—like an identifier, punctuation, or a literal—or another TokenStream enclosed in a delimiter like () or {}. By walking a TokenStream, you can parse out whatever syntax you wish as long as the individual tokens are valid Rust tokens. If you want to parse your input specifically as Rust code, you will likely want to use the syn crate, which implements a complete Rust parser and can turn a TokenStream into an easy-to-traverse Rust AST.

With most procedural macros, you want to not only parse a TokenStream but also produce Rust code to be injected into the program that invokes the procedural macro. There are two main ways to do so. The first is to manually construct a TokenStream and extend it one TokenTree at a time. The second is to use TokenStream's implementation of FromStr, which lets you parse a string that contains Rust code into a TokenStream with "".parse::<TokenStream>(). You can also mix and match these; if you want to prepend some code to your macro's input, just construct a TokenStream for the prologue, and then use the Extend trait to append the original input.

NOTE   *TokenStream also implements Display, which pretty-prints the tokens in the stream. This comes in super handy for debugging!*

Tokens are very slightly more magical than I've described so far in that every token, and indeed every TokenTree, also has a *span*. Spans are how the compiler ties generated code back to the source code that generated it. Every token's span marks where that token originated. For example, consider a (declarative) macro like the one in Listing 7-7, which generates a trivial Debug implementation for the provided type.

```
macro_rules! name_as_debug {
  ($t:ty) => {
    impl ::core::fmt::Debug for $t {
      fn fmt(&self, f: &mut ::core::fmt::Formatter<'_>) -> ::core::fmt::Result
      { ::core::write!(f, ::core::stringify!($t)) }
} }; }
```

Listing 7-7: A very simple macro for implementing Debug

Now let's imagine that someone invokes this macro with name_as_debug!(u31). Technically, the compiler error occurs inside the macro, specifically where we write for $t (the other use of $t can handle an invalid type). But we'd like the compiler to point the user at the u31 in their code—and indeed, that's what spans let us do.

The span of the $t in the generated code is the code mapped to $t in the macro invocation. That information is then carried through the compiler and associated with the eventual compiler error. When that compiler error is eventually printed, the compiler will print the error from inside the macro saying that the type u31 does not exist but will highlight the u31 argument in the macro invocation, since that's the error's associated span!

Spans are quite flexible, and they enable you to write procedural macros that can produce sophisticated error messages if you use the compile_error! macro. As its name implies, compile_error! causes the compiler to emit an error wherever it is placed with the provided string as the message. This may not seem very useful, until you pair it with a span. By setting the span of the TokenTree you generate for the compile_error! invocation to be equal to the span of some subset of the input, you are effectively telling the compiler to emit this compiler error and point the user to this part of the source. Together, these two mechanisms let a macro produce errors that seem to stem from the relevant part of the code, even though the actual compiler error is somewhere in the generated code that the user never even sees!

NOTE *If you've ever been curious how* syn*'s error handling works, its* Error *type imple-ments an* Error::to_compile_error *method, which turns it into a* TokenStream *that holds only a* compile_error! *directive. What's particularly neat with* syn*'s* Error *type is that it internally holds a collection of errors, each of which produces a distinct* compile_error! *directive with its own span so that you can easily produce multiple independent errors from your procedural macro.*

The power of spans doesn't end there; spans are also how Rust's macro hygiene is implemented. When you construct an Ident token, you also give the span for that identifier, and that span dictates the scope of that identi-fier. If you set the identifier's span to be Span::call_site(), the identifier is resolved where the macro was called from and will thus not be isolated from the surrounding scope. If, on the other hand, you set it to Span::mixed _site() then (variable) identifiers are resolved at the macro definition site, and so will be completely hygienic with respect to similarly named variables at the call site. Span::mixed_site is so called because it matches the rules around identifier hygiene for macro_rules!, which, as we discussed earlier, "mixes" identifier resolution between using the macro definition site for variables and using the call site for types, modules, and everything else.

Summary

In this chapter we covered both declarative and procedural macros, and looked at when you might find each of them useful in your own code. We also took a deeper dive into the mechanisms that underpin each type of macro and some of the features and gotchas to be aware of when you write your own macros. In the next chapter, we'll start our journey into asyn-chronous programming and the Future trait. I promise—it's just on the next page.

8

ASYNCHRONOUS PROGRAMMING

Asynchronous programming is, as the name implies, programming that is not synchronous. At a high level, an asynchronous operation is one that executes in the background—the program won't wait for the asynchronous operation to complete but will instead continue to the next line of code immediately.

If you're not already familiar with asynchronous programming, that definition may feel insufficient as it doesn't actually explain what asynchronous programming *is*. To really understand the asynchronous programming model and how it works in Rust, we have to first dig into what the alternative is. That is, we need to understand the *synchronous* programming model before we can understand the *asynchronous* one. This is important in both clarifying the concepts and demonstrating the trade-offs of using asynchronous programming: an asynchronous solution is not always the right one! We'll start this chapter by taking a quick journey through what motivates asynchronous programming as a concept in the first place; then we'll dig into how asynchrony in Rust actually works under the hood.

What's the Deal with Asynchrony?

Before we get to the details of the synchronous and asynchronous programming models, we first need to take a quick look at what your computer is actually doing when it runs your programs.

Computers are fast. Really fast. So fast, in fact, that they spend most of their time waiting for things to happen. Unless you're decompressing files, encoding audio, or crunching numbers, chances are that your CPU mostly sits idle, waiting for operations to complete. It's waiting for a network packet to arrive, for the mouse to move, for the disk to finish writing some bytes, or maybe even just for a read from main memory to complete. From the CPU's perspective, eons go by between most such events. When one does occur, the CPU runs a few more instructions, then goes back to waiting again. Take a look at your CPU utilization—it's probably somewhere in the low single digits, and that's likely where it hovers the majority of the time.

Synchronous Interfaces

Synchronous interfaces allow your program (or rather, a single thread in your program) to execute only a single operation at a time; each operation has to wait for the previous synchronous operation to finish before it gets to run. Most interfaces you see in the wild are synchronous: you call them, they go do some stuff, and eventually they return when the operation has completed and your program can continue from there. The reason for this, as we'll see later in this chapter, is that making an operation asynchronous takes a fair bit of extra machinery. Unless you need the benefits of asynchrony, sticking to the synchronous model requires much less pomp and circumstance.

Synchronous interfaces hide all this waiting; the application calls a function that says "write these bytes to this file," and some time later, that function completes and the next line of code executes. Behind the scenes, what really happens is that the operating system queues up a write operation to the disk and then puts the application to sleep until the disk reports that it has finished the write. The application experiences this as the function taking a long time to execute, but in reality it isn't really executing at all, just waiting.

An interface that performs operations sequentially in this way is also often referred to as *blocking*, since the operation in the interface that has to wait for some external event to happen in order for it to make progress *blocks* further execution until that event happens. Whether you refer to an interface as synchronous or blocking, the basic idea is the same: the application does not move on until the current operation finishes. While the operation is waiting, so is the application.

Synchronous interfaces are usually considered to be easy to work with and simple to reason about, since your code executes just one line at a time.

But they also allow the application to do only one thing at a time. That means if you want your program to wait for either user input or a network packet, you're out of luck unless your operating system provides an operation specifically for that. Similarly, even if your application could do some other useful work while the disk is writing a file, it doesn't have that option as the file write operation blocks the execution!

Multithreading

By far the most common solution to allowing concurrent execution is to use *multithreading*. In a multithreaded program, each thread is responsible for executing a particular independent sequence of blocking operations, and the operating system multiplexes among the threads so that if any thread can make progress, progress is made. If one thread blocks, some other thread may still be runnable, and so the application can continue to do useful work.

Usually, these threads communicate with each other using a synchronization primitive like a lock or a channel so that the application can still coordinate their efforts. For example, you might have one thread that waits for user input, one thread that waits for network packets, and another thread that waits for either of those threads to send a message on a channel shared between all three threads.

Multithreading gives you *concurrency*—the ability to have multiple independent operations that can be executed at any one time. It's up to the system running the application (in this case, the operating system) to choose among the threads that aren't blocked and decide which to execute next. If one thread is blocked, it can choose to run another one that can make progress instead.

Multithreading combined with blocking interfaces gets you quite far, and large swaths of production-ready software are built in this way. But this approach is not without its shortcomings. First, keeping track of all these threads quickly gets cumbersome; if you have to spin up a thread for every concurrent task, including simple ones like waiting for keyboard input, the threads add up fast, and so does the additional complexity needed to keep track of how all those threads interact, communicate, and coordinate.

Second, switching between threads gets costly the more of them there are. Every time one thread stops running and another one starts back up in its place, you need to do a round-trip to the operating system scheduler, and that's not free. On some platforms, spawning new threads is also a fairly heavyweight process. Applications with high performance needs often mitigate this cost by reusing threads and using operating system calls that allow you to block on many related operations, but ultimately you are left with the same problem: blocking interfaces require that you have as many threads as the number of blocking calls you want to make.

Finally, threads introduce *parallelism* into your program. The distinction between concurrency and parallelism is subtle, but important: concurrency means that the execution of your tasks is interleaved, whereas parallelism means that multiple tasks are executing at the same time. If you have two tasks, their execution expressed in ASCII might look like _-_-_

(concurrency) versus ===== (parallelism). Multithreading does not necessarily imply parallelism—even though you have many threads, you might have only a single core, so only one thread is executing at a given time—but the two usually go hand in hand. You can make two threads mutually exclusive in their execution by using a `Mutex` or other synchronization primitive, but that introduces additional complexity—threads want to run in parallel. And while parallelism is often a good thing—who doesn't want their program to run faster on more cores—it also means that your program must handle truly simultaneous access to shared data structures. This means moving from `Rc`, `Cell`, and `RefCell` to the more powerful but also slower `Arc` and `Mutex`. While you *may* want to use the latter types in your concurrent program to enable parallelism, threading *forces* you to use them. We'll look at multithreading in much greater detail in Chapter 10.

Asynchronous Interfaces

Now that we've explored synchronous interfaces, we can look at the alternative: asynchronous or *nonblocking* interfaces. An asynchronous interface is one that may not yield a result straightaway, and may instead indicate that the result will be available at some later time. This gives the caller the opportunity to do something else in the meantime rather than having to go to sleep until that particular operation completes. In Rust parlance, an asynchronous interface is a method that returns a `Poll`, as defined in Listing 8-1.

```
enum Poll<T> {
    Ready(T),
    Pending
}
```

Listing 8-1: The core of asynchrony: the "here you are or come back later" type

`Poll` usually shows up in the return type of functions whose names start with `poll`—these are methods that signal they can attempt an operation without blocking. We'll get into how exactly they do that later in this chapter, but in general they attempt to perform as much as they can of the operation before they would normally block, and then return. And crucially, they remember where they left off so that they can resume execution later when additional progress can again be made.

These nonblocking functions allow us to easily perform multiple tasks concurrently. For example, if you want to read from either the network or the user's keyboard, whichever has an event available first, all you have to do is poll both in a loop until one of them returns `Poll::Ready`. No need for any additional threads or synchronization!

The word *loop* here should make you a little nervous. You don't want your program to burn through a loop three billion times a second when it may be minutes until the next input occurs. In the world of blocking interfaces, this wasn't a problem since the operating system simply put the thread to sleep and then took care of waking it up when a relevant event occurred, but how do we avoid burning cycles while waiting in this brave new nonblocking world? That's what much of the remainder of this chapter will be about.

Standardized Polling

To get to a world where every library can be used in a nonblocking fashion, we could have every library author cook up their own `poll` methods, all with slightly different names, signatures, and return types—but that would quickly get unwieldy. Instead, in Rust, polling is standardized through the `Future` trait. A simplified version of `Future` is shown in Listing 8-2 (we'll get back to the real one later in this chapter).

```
trait Future {
    type Output;
    fn poll(&mut self) -> Poll<Self::Output>;
}
```

Listing 8-2: A simplified view of the `Future` trait

Types that implement the `Future` trait are known as *futures* and represent values that may not be available yet. A future could represent the next time a network packet comes in, the next time the mouse cursor moves, or just the point at which some amount of time has elapsed. You can read `Future<Output = Foo>` as "a type that will produce a `Foo` in the future." Types like this are often referred to in other languages as *promises*—they promise that they will eventually yield the indicated type. When a future eventually returns `Poll::Ready(T)`, we say that the future *resolves* into a `T`.

With this trait in place, we can generalize the pattern of providing `poll` methods. Instead of having methods like `poll_recv` and `poll_keypress`, we can have methods like `recv` and `keypress` that both return `impl Future` with an appropriate `Output` type. This doesn't change the fact that you have to poll them—we'll deal with that later—but it does mean that at least there is a standardized interface to these kinds of pending values, and we don't need to use the `poll_` prefix everywhere.

NOTE *In general, you should not poll a future again after it has returned `Poll::Ready`. If you do, the future is well within its rights to panic. A future that is safe to poll after it has returned `Ready` is sometimes referred to as a* fused *future.*

Ergonomic Futures

Writing a type that implements `Future` in the way I've described so far is quite a pain. To see why, first take a look at the fairly straightforward asynchronous code block in Listing 8-3 that simply tries to forward messages from the input channel rx to the output channel tx.

```
async fn forward<T>(rx: Receiver<T>, tx: Sender<T>) {
    while let Some(t) = rx.next().await {
        tx.send(t).await;
    }
}
```

Listing 8-3: Implementing a channel-forwarding future using async and await

This code, written using async and await syntax, looks very similar to its equivalent synchronous code and is easy to read. We simply send each message we receive in a loop until there are no more messages, and each await point corresponds to a place where a synchronous variant might block. Now think about if you instead had to express this code by manually implementing the Future trait. Since each call to poll starts at the top of the function, you'd need to package the necessary state to continue from the last place the code yielded. The result is fairly grotesque, as Listing 8-4 demonstrates.

```rust
enum Forward<T> { ❶
    WaitingForReceive(ReceiveFuture<T>, Option<Sender<T>>),
    WaitingForSend(SendFuture<T>, Option<Receiver<T>>),
}

impl<T> Future for Forward<T> {
    type Output = (); ❷
    fn poll(&mut self) -> Poll<Self::Output> {
        match self { ❸
            Forward::WaitingForReceive(recv, tx) => {
                if let Poll::Ready((rx, v)) = recv.poll() {
                    if let Some(v) = v {
                        let tx = tx.take().unwrap(); ❹
                        *self = Forward::WaitingForSend(tx.send(v), Some(rx)); ❺
                        // Try to make progress on sending.
                        return self.poll(); ❻
                    } else {
                        // No more items.
                        Poll::Ready(())
                    }
                } else {
                    Poll::Pending
                }
            }
            Forward::WaitingForSend(send, rx) => {
                if let Poll::Ready(tx) = send.poll() {
                    let rx = rx.take().unwrap();
                    *self = Forward::WaitingForReceive(rx.receive(), Some(tx));
                    // Try to make progress on receiving.
                    return self.poll();
                } else {
                    Poll::Pending
                }
            }
        }
    }
}
```

Listing 8-4: Manually implementing a channel-forwarding future

You'll rarely have to write code like this in Rust anymore, but it gives important insight into how things work under the hood, so let's walk through it. First, we define our future type as an enum ❶, which we'll use to keep track of what we're currently waiting on. This is a consequence of the fact that when we return `Poll::Pending`, the next call to `poll` will start at the top of the function again. We need some way to know what we were in the middle of so that we know which operation to continue on. Furthermore, we need to keep track of different information depending on what we're doing: if we're waiting for a receive to finish, we need to keep that `ReceiveFuture` (the definition of which is not shown in this example) so that we can poll it the next time we are polled ourselves, and the same goes for `SendFuture`. The `Options` here might strike you as weird too; we'll get back to those shortly.

When we implement `Future` for `Forward`, we declare its output type as () ❷ because this future doesn't actually return anything. Instead, the future resolves (with no result) when it has finished forwarding everything from the input channel to the output channel. In a more complete example, the `Output` of our forwarding type might be a `Result` so that it could communicate errors from `receive()` and `send()` back up the stack to the function that's polling for the completion of the forwarding. But this code is complicated enough already, so we'll leave that for another day.

When `Forward` is polled, it needs to resume wherever it last left off, which we find out by matching on the enum variant currently held in `self` ❸. For whichever branch we go into, the first step is to poll the future that blocks progress for the current operation; if we're trying to receive, we poll the `ReceiveFuture`, and if we're trying to send, we poll the `SendFuture`. If that call to `poll` returns `Poll::Pending`, then we can make no progress, and we return `Poll::Pending` ourselves. But if the current future resolves, we have work to do!

When one of the inner futures resolves, we need to update what the current operation is by switching which enum variant is stored in `self`. In order to do so, we have to move out of `self` to call `Receiver::receive` or `Sender::send`—but we can't do that because all we have is `&mut self`. So, we store the state we have to move in an `Option`, which we move out of with `Option::take` ❹. This is silly since we're about to overwrite `self` anyway ❺, and hence the `Options` will always be `Some`, but sometimes tricks are needed to make the borrow checker happy.

Finally, if we do make progress, we then poll `self` again ❻ so that if we can immediately make progress on the pending send or receive, we do so. This is actually necessary for correctness when implementing the real `Future` trait, which we'll get back to later, but for now think of this as an optimization.

We just hand-wrote a *state machine*: a type that has a number of possible states and moves between them in response to particular events. This was a fairly simple state machine, at that. Imagine having to write code like this for more complicated use cases where you have additional intermediate steps!

Beyond writing the unwieldy state machine, we have to know the types of the futures that `Sender::send` and `Receiver::receive` return so that we can store them in our type. If those methods instead returned `impl Future`, we'd have no way to write out the types for our variants. The `send` and `receive` methods also have to take ownership of the sender and the receiver; if they did not, the lifetimes of the futures they returned would be tied to the borrow of self, which would end when we return from poll. But that would not work, since we're trying to store those futures *in* self.

NOTE   *You may have noticed that Receiver looks a lot like an asynchronous version of Iterator. Others have noticed the same thing, and the standard library is on its way to adding a trait specifically for types that can meaningfully implement poll_next. Down the line, these asynchronous iterators (often referred to as streams) may end up with first-class language support, such as the ability to loop over them directly!*

Ultimately, this code is hard to write, hard to read, and hard to change. If we wanted to add error handling, for example, the code complexity would increase significantly. Luckily, there's a better way!

async/await

Rust 1.39 gave us the async keyword and the closely related await postfix operator, which we used in the original example in Listing 8-3. Together, they provide a much more convenient mechanism for writing asynchronous state machines like the one in Listing 8-5. Specifically, they let you write the code in such a way that it doesn't even look like a state machine!

```
async fn forward<T>(rx: Receiver<T>, tx: Sender<T>) {
    while let Some(t) = rx.next().await {
        tx.send(t).await;
    }
}
```

Listing 8-5: Implementing a channel-forwarding future using async and await, repeated from Listing 8-3

If you don't have much experience with async and await, the difference between Listing 8-4 and Listing 8-5 might give you an idea of why the Rust community was so excited to see them land. But since this is an intermediate book, let's dive a little deeper to understand just how this short segment of code can replace the much longer manual implementation. To do that, we first need to talk about *generators*—the mechanism by which async and await are implemented.

Generators

Briefly described, a generator is a chunk of code with some extra compiler-generated bits that enables it to stop, or *yield*, its execution midway through and then resume from where it last yielded later on. Take the forward function in Listing 8-3, for example. Imagine that it gets to the call to send, but the channel is currently full. The function can't make any more progress, but it also cannot block (this is nonblocking code, after all), so it needs to return. Now suppose the channel eventually clears and we want to proceed with the send. If we call forward again from the top, it'll call next again and the item we previously tried to send will be lost, so that's no good. Instead, we turn forward into a generator.

Whenever the forward generator cannot make progress anymore, it needs to store its current state somewhere so that when its execution eventually resumes, it resumes in the right place with the right state. It saves the state through an associated data structure that's generated by the compiler, which contains all the state of the generator at a given point in time. A method on that data structure (also generated) then allows the generator to resume from its current state, stored in &mut self, and updates the state again when the generator again cannot make progress.

This "return but allow me to resume later" operation is called *yielding*, which effectively means it returns while keeping some extra state on the side. When we later want to resume a call to forward, we invoke the known entry point into the generator (the *resume method*, which is poll for async generators), and the generator inspects the previously stored state in self to decide what to do next. This is exactly the same thing we did manually in Listing 8-4! In other words, the code in Listing 8-5 loosely desugars to the hypothetical code shown in Listing 8-6.

```
generator fn forward<T>(rx: Receiver<T>, tx: Sender<T>) {
    loop {
        let mut f = rx.next();
        let r = if let Poll::Ready(r) = f.poll() { r } else { yield };
        if let Some(t) = r {
            let mut f = tx.send(t);
            let _ = if let Poll::Ready(r) = f.poll() { r } else { yield };
        } else { break Poll::Ready(()); }
    }
}
```

Listing 8-6: Desugaring async/await into a generator

At the time of writing, generators are not actually usable in Rust—they are only used internally by the compiler to implement async/await—but that may change in the future. Generators come in handy in a number of cases, such as to implement iterators without having to carry around a struct or to implement an impl Iterator that figures out how to yield items one at a time.

If you look closely at Listings 8-5 and 8-6, they may seem a little magical once you know that every await or yield is really a return from the function. After all, there are several local variables in the function, and it's not clear how they're restored when we resume later on. This is where the compiler-generated part of generators comes into play. The compiler transparently injects code to persist those variables into and read them from the generator's associated data structure, rather than the stack, at the time of execution. So if you declare, write to, or read from some local variable a, you are really operating on something akin to self.a. Problem solved! It's all really quite marvelous.

One subtle but important difference between the manual forward implementation and the async/await version is that the latter can hold references across yield points. This enables functions like Receiver::next and Sender::send in Listing 8-5 to take &mut self rather than the self they took in Listing 8-4. If we tried to use a &mut self receiver for these methods in the manual state machine implementation, the borrow checker would have no way to enforce that the Receiver stored inside Forward cannot be referenced between when Receiver::next is called and when the future it returns resolves, and so it would reject the code. Only by moving the Receiver into the future can we convince the compiler that the Receiver is not otherwise accessible. Meanwhile, with async/await, the borrow checker can inspect the code before the compiler turns it into a state machine and verify that rx is indeed not accessed again until after the future is dropped, when the await on it returns.

THE SIZE OF GENERATORS

The data structure used to back a generator's state must be able to hold the combined state at any one yield point. If your async fn contains, say, a [u8; 8192], those 8KiB must be stored in the generator itself. Even if your async fn contains only smaller local variables, it must also contain any future that it awaits, since it needs to be able to poll such a future later, when poll is invoked.

This nesting means that generators, and thus futures based on async functions and blocks, can get quite large without any visible indicator of that increased size in your code. This can in turn impact your program's runtime performance, since those giant generators may have to be copied across function calls and in and out of data structures, which amounts to a fair amount of memory copying. In fact, you can usually identify when the size of your generator-based futures is affecting performance by looking for excessive amounts of time spent in the memcpy function in your application's performance profiles!

Finding these large futures isn't always easy, however, and often requires manually identifying long or complex chains of async functions. Clippy may be able to help with this in the future, but at the time of writing, you're on your own. When you do find a particularly large future, you have two options: you can try to reduce the amount of local state the async functions need, or you can move the future to the heap (with Box::pin) so that moving the future just requires moving the pointer to it. The latter is by far the easiest to go, but it also introduces an extra allocation and a pointer indirection. Your best bet is usually to put the problematic future on the heap, measure your performance, and then use your performance benchmarks to guide you from there.

Pin and Unpin

We're not quite done. While generators are neat, a challenge arises from the technique as I've described it so far. In particular, it's not clear what happens if the code in the generator (or, equivalently, the async block) takes a reference to a local variable. In the code from Listing 8-5, the future that rx.next() returns must necessarily hold a reference to rx if a next message is not immediately available so that it knows where to try again when the generator next resumes. When the generator yields, the future and the reference the future contains get stashed away inside the generator. But what now happens if the generator is moved? Specifically, look at the code in Listing 8-7, which calls forward.

```
async fn try_forward<T>(rx: Receiver<T>, tx: Sender<T>) -> Option<impl Future>
{
    let mut f = forward(rx, tx);
    if f.poll().is_pending() { Some(f) } else { None }
}
```

Listing 8-7: Moving a future after polling it

The try_forward function polls forward only once, to forward as many messages as possible without blocking. If the receiver may still produce more messages (that is, if it returned Poll::Pending instead of Poll::Ready(None)), those messages are deferred to be forwarded at some later time by returning the forwarding future to the caller, which may choose to poll again at a time when it sees fit.

Let's work through what happens here with what we know about async and await so far. When we poll the forward generator, it goes through the while loop some unknown number of times and eventually returns either Poll::Ready(()) if the receiver ended, or Poll::Pending otherwise. If it returns Poll::Pending, the generator contains a future returned from either rx.next() or tx.send(t). Those futures both contain a reference to one of the arguments initially provided to forward (rx and tx, respectively), which must also be stored in the generator. But when try_forward returns the entire generator, the fields of the generator also move. Thus, rx and tx no longer reside at the same locations in memory, and the references stored in the stashed-away future are no longer pointing to the right data!

What we've run into here is a case of a *self-referential* data structure: one that holds both data and references to that data. With generators, these self-referential structures are very easy to construct, and being unable to support them would be a significant blow to ergonomics because it would mean you wouldn't be able to hold references across any yield point. The (ingenious) solution for supporting self-referential data structures in Rust comes in the form of the Pin type and the Unpin trait. Very briefly, Pin is a wrapper type that prevents the wrapped type from being (safely) moved, and Unpin is a marker trait that says the implementing type *can* be removed safely from a Pin.

Pin

There's a lot of nuance to cover here, so let's start with a concrete use of the Pin wrapper. Listing 8-2 gave you a simplified version of the Future trait, but we're now ready to peel back one part of the simplification. Listing 8-8 shows the Future trait somewhat closer to its final form.

```
trait Future {
    type Output;
    fn poll(self: Pin<&mut Self>) -> Poll<Self::Output>;
}
```

Listing 8-8: A less simplified view of the Future trait with Pin

In particular, this definition requires that you call `poll` on `Pin<&mut Self>`. Once you have a value behind a `Pin`, that constitutes a contract that that value will never move again. This means that you can construct self-references internally to your heart's delight, exactly as you want for generators.

While Future *makes use of* Pin, Pin *is not tied to the* Future *trait—you can use* Pin *for any self-referential data structure.*

But how do you get a `Pin` to call `poll`? And how can `Pin` ensure that the contained value won't move? To see how this magic works, let's look at the definition of `std::pin::Pin` and some of its key methods, shown in Listing 8-9.

```
struct Pin<P> { pointer: P }
impl<P> Pin<P> where P: Deref {
    pub unsafe fn new_unchecked(pointer: P) -> Self;
}
impl<'a, T> Pin<&'a mut T> {
    pub unsafe fn get_unchecked_mut(self) -> &'a mut T;
}
impl<P> Deref for Pin<P> where P: Deref {
    type Target = P::Target;
    fn deref(&self) -> &Self::Target;
}
```

Listing 8-9: std::pin::Pin and its key methods

There's a lot to unpack here, and we're going to have to go over the definition in Listing 8-9 a few times before all the bits make sense, so please bear with me.

First, you'll notice that `Pin` holds a *pointer type*. That is, rather than hold some `T` directly, it holds a type `P` that dereferences through `Deref` into `T`. This means that rather than have a `Pin<MyType>`, you'll have a `Pin<Box<MyType>>` or `Pin<Rc<MyType>>` or `Pin<&mut MyType>`. The reason for this design is simple— `Pin`'s primary goal is to make sure that once you place a `T` behind a `Pin`, that `T` won't move, as doing so might invalidate self-references stored in the `T`. If the `Pin` just held a `T` directly, then simply moving the `Pin` would be enough to invalidate that invariant! In the remainder of this section, I'll refer to `P` as the *pointer* type and `T` as the *target* type.

Next, notice that `Pin`'s constructor, `new_unchecked`, is unsafe. This is because the compiler has no way to actually check that the pointer type indeed promises that the pointed-to (target) type won't move again. Consider, for example, a variable `foo` on the stack. If `Pin`'s constructor were safe, we could do `Pin::new(&mut foo)`, call a method that requires `Pin<&mut Self>` (and thus assumes that `Self` won't move again), and then drop the `Pin`. At this point, we could modify `foo` as much as we liked, since it is no longer borrowed—including moving it! We could then pin it again and call the same method, which would be none the wiser that any self-referential pointers it may have constructed the first time around would now be invalid.

We then get to the get_unchecked_mut method, which gives you a mutable reference to the T behind the Pin's pointer type. This method is also unsafe, because once we give out a &mut T, the caller has to promise it won't use that &mut T to move the T or otherwise invalidate its memory, lest any self-references be invalidated. If this method weren't unsafe, a caller could call a method that takes Pin<&mut Self> and then call the safe variant of get_unchecked_mut on two Pin<&mut _>s, then use mem::swap to swap the values behind the Pin. If we were to then call a method that takes Pin<&mut Self> again on either Pin, its assumption that the Self hasn't moved would be violated, and any internal references it stored would be invalid!

Perhaps surprisingly, Pin<P> always implements Deref<Target = T>, and that is entirely safe. The reason for this is that a &T does not let you move T without writing other unsafe code (UnsafeCell, for example, as we'll discuss in Chapter 9). This is a good example of why the scope of an unsafe block extends beyond just the code it contains. If you wrote some code in one part of the application that (unsafely) replaced a T behind an & using UnsafeCell, then it *could* be that that &T initially came from a Pin<&mut T>, and that you have now violated the invariant that the T behind the Pin may never move, even though the place where you unsafely replaced the &T did not even mention Pin!

NOTE *If you've browsed through the Pin documentation while reading this chapter, you may have noticed Pin::set, which takes a &mut self and a <P as Deref>::Target and safely changes the value behind the Pin. This is possible because set does not return the value that was previously pinned—it simply drops it in place and stores the new value there instead. Therefore, it does not violate the pinning invariants: the old value was never accessed outside of a Pin after it was placed there.*

Unpin: The Key to Safe Pinning

At this point you might ask: given that getting a mutable reference is unsafe anyway, why not have Pin hold a T directly? That is, rather than require an indirection through a pointer type, you could instead make the contract for get_unchecked_mut that it is only safe to call if you haven't moved the Pin. The answer to that question lies in a neat safe use of Pin that the pointer design enables. Recall that the whole reason we want Pin in the first place is so we can have target types that may contain references to themselves (like a generator) and give their methods a guarantee that the target type hasn't moved and thus that internal self-references remain valid. Pin lets us use the type system to enforce that guarantee, which is great. But unfortunately, with the design so far, Pin is very unwieldy to work with. This is because it always requires unsafe code, even if you are working with a target type that doesn't contain any self-references, and so doesn't care whether it's been moved or not.

This is where the marker trait Unpin comes into play. An implementation of Unpin for a type simply asserts that the type is safe to move out of a Pin when used as a target type. That is, the type promises that it will never use any of Pin's guarantees about the referent not moving again when used as a target type, and thus those guarantees may be broken. Unpin is an auto-trait, like Send and Sync, and so is auto-implemented by the compiler for any type that contains only Unpin members. Only types that explicitly opt out of Unpin (like generators) and types that contain those types are !Unpin.

For target types that are Unpin, we can provide a much simpler safe interface to Pin, as shown in Listing 8-10.

```
impl<P> Pin<P> where P: Deref, P::Target: Unpin {
    pub fn new(pointer: P) -> Self;
}
impl<P> DerefMut for Pin<P> where P: DerefMut, P::Target: Unpin {
    fn deref_mut(&mut self) -> &mut Self::Target;
}
```

Listing 8-10: The safe API to Pin for Unpin target types

To make sense of the safe API in Listing 8-10, think about the safety requirements of the unsafe methods from Listing 8-9: the function Pin::new_unchecked is unsafe because the caller must promise that the referent cannot be moved outside of the Pin, and that the implementations of Deref, DerefMut, and Drop for the pointer type do not move the referent through the reference they receive. Those requirements are there to ensure that once we give out a Pin to a T, we never move that T again. But if the T is Unpin, it has declared that it does not care if it is moved even if it was previously pinned, so it's fine if the caller does not satisfy any of those requirements!

Similarly, get_unchecked_mut is unsafe because the caller must guarantee that it doesn't move the T out of the &mut T—but with T: Unpin, T has declared that it's fine being moved even after being pinned, so that safety requirement is no longer important. This means that for Pin<P> where P::Target: Unpin, we can simply provide safe variants of both those methods (DerefMut being the safe version of get_unchecked_mut). In fact, we can even provide a Pin::into_inner that simply gives back the owned P if the target type is Unpin, since the Pin is essentially irrelevant!

Ways of Obtaining a Pin

With our new understanding of `Pin` and `Unpin`, we can now make progress toward using the new `Future` definition from Listing 8-8 that requires `Pin<&mut Self>`. The first step is to construct the required type. If the future type is `Unpin`, that step is easy—we just use `Pin::new(&mut future)`. If it is not `Unpin`, we can pin the future in one of two main ways: by pinning to the heap or pinning to the stack.

Let's start with pinning to the heap. The primary contract of `Pin` is that once something has been pinned, it cannot move. The pinning API takes care of honoring that contract for all methods and traits on `Pin`, so the main role of any function that constructs a `Pin` is to ensure that if the `Pin` *itself* moves, the referent value does not move too. The easiest way to ensure that is to place the referent on the heap, and then place just a pointer to the referent in the `Pin`. You can then move the `Pin` to your heart's delight, but the target will remain where it was. This is the rationale behind the (safe) method `Box::pin`, which takes a `T` and returns a `Pin<Box<T>>`. There's no magic to it; it simply asserts that `Box` follows the `Pin` constructor, `Deref`, and `Drop` contracts.

UNPIN BOX

While we're on the topic of `Box`, take a look at the implementation of `Unpin` for `Box`. The `Box` type unconditionally implements `Unpin` for any `T`, even if that `T` is not `Unpin`. This might strike you as odd, given the earlier assertion that `Unpin` is an auto-trait that is generally implemented for a type only if all of the type's members are also `Unpin`. `Box` is an exception to this for the same reason that it can provide a safe `Pin` constructor: if you move a `Box<T>`, you do not move the `T`. In other words, the unconditional implementation asserts that you can move a `Box<T>` out of a `Pin` even if `T` cannot be moved out of a `Pin`. Note, however, that this does *not* enable you to move a `T` that is `!Unpin` out of a `Pin<Box<T>>`.

The other option, pinning to the stack, is a little more involved, and at the time of writing requires a smidgen of unsafe code. We have to ensure that the pinned value cannot be accessed after the `Pin` with a `&mut` to it has been dropped. We accomplish that by shadowing the value as shown in the macro in Listing 8-11 or by using one of the crates that provide exactly this macro. One day it may even make it into the standard library!

```
macro_rules! pin_mut {
    ($var:ident) => {
        let mut $var = $var;
        let mut $var = unsafe { Pin::new_unchecked(&mut $var) };
    }
}
```

Listing 8-11: Macro for pinning to the stack

By taking the name of the variable to pin to the stack, the macro ensures that the caller has the value it wants to pin somewhere on the stack already. The shadowing of $var ensures that the caller cannot drop the Pin and continue to use the unpinned value (which would breach the Pin contract for any target type that's !Unpin). By moving the value stored in $var, the macro also ensures that the caller cannot drop the $var binding the macro declarations without also dropping the original variable. Specifically, without that line, the caller could write (note the extra scope):

```
let foo = /* */; { pin_mut!(foo); foo.poll() }; foo.mut_self_method();
```

Here, we give a pinned instance of foo to poll, but then we later use a &mut to foo without a Pin, which violates the Pin contract. With the extra reassignment, on the other hand, that code would also move foo into the new scope, rendering it unusable after the scope ends.

Pinning on the stack therefore requires unsafe code, unlike Box::pin, but avoids the extra allocation that Box introduces and also works in no_std environments.

Back to the Future

We now have our pinned future, and we know what that means. But you may have noticed that none of this important pinning stuff shows up in most asynchronous code you write with async and await. And that's because the compiler hides it from you.

Think back to when we discussed Listing 8-5, when I told you that <expr>.await desugars into something like:

```
loop { if let Poll::Ready(r) = expr.poll() { break r } else { yield } }
```

That was an ever-so-slight simplification because, as we've seen, you can call Future::poll only if you have a Pin<&mut Self> for the future. The desugaring is actually a bit more sophisticated, as shown in Listing 8-12.

```
❶ match expr {
    mut pinned => loop {
      ❷ match unsafe { Pin::new_unchecked(&mut pinned) }.poll() {
          Poll::Ready(r) => break r,
          Poll::Pending => yield,
      }
    }
}
```

Listing 8-12: A more accurate desugaring of <expr>.await

The match ❶ is a neat shorthand to not only ensure that the expansion remains a valid expression, but also move the expression result into a variable that we can then pin on the stack. Beyond that, the main new addition is the call to Pin::new_unchecked ❷. That call is safe because for the containing async block to be polled, it must already be pinned due to the signature of Future::poll. And the async block was polled for us to reach the call to `Pin::new_unchecked`, so the generator state is pinned. Since `pinned` is stored in the generator that corresponds to the async block (it must be so that `yield` will resume correctly), we know that `pinned` will not move again. Furthermore, `pinned` is not accessible except through a `Pin` once we're in the loop, so no code is able to move out of the value in `pinned`. Thus, we meet all the safety requirements of `Pin::new_unchecked`, and the code is safe.

Going to Sleep

We went pretty deep into the weeds with `Pin`, but now that we're out the other side, there is another issue around futures that may have been making your brain itch. If a call to `Future::poll` returns `Poll::Pending`, you need something to call `poll` again at a later time to check whether you can make progress yet. That something is usually called the *executor*. Your executor could be a simple loop that polls all the futures you are waiting on until they've all returned `Poll::Ready`, but that would burn a lot of CPU cycles you could probably have used for other, more useful things, like running your web browser. Instead, we want the executor to do whatever useful work it can do, and then go to sleep. It should stay asleep until one of the futures can make progress, and only then wake up to do another pass, before going to sleep again.

Waking Up

The condition that determines when to check back with a given future varies widely. It might be "when a network packet arrives on this port," "when the mouse cursor moves," "when someone sends on this channel," "when the CPU receives a particular interrupt," or even "after this much time has passed." On top of that, developers can write their own futures that wrap multiple other futures, and thus, they may have several wake-up conditions. Some futures may even introduce their own entirely custom wake events.

To accommodate these many use cases, Rust introduces the notion of a `Waker`: a way to wake the executor to signal that progress can be made. The `Waker` is what makes the whole machinery around futures work. The executor constructs a `Waker` that integrates with the mechanism the executor uses to go to sleep, and passes the `Waker` in to any `Future` it polls. How? With the additional parameter to `Future::poll` that I've hidden from you so far. Sorry about that. Listing 8-13 gives the final and true definition for `Future`—no more lies!

```
trait Future {
    type Output;
    fn poll(self: Pin<&mut Self>, cx: &mut Context<'_>) -> Poll<Self::Output>;
}
```

Listing 8-13: The actual Future trait with Context

The `&mut Context` contains the `Waker`. The argument is a `Context`, not a `Waker` directly, so that we can augment the asynchronous ecosystem with additional context for futures should that be deemed necessary.

The primary method on Waker is wake (and the by-reference variant wake _by_ref), which should be called when the future can again make progress. The wake method takes no arguments, and its effects are entirely defined by the executor that constructed the Waker. You see, Waker is secretly generic over the executor. Or, more precisely, whatever constructed the Waker gets to dictate what happens when Waker::wake is called, when a Waker is cloned, and when a Waker is dropped. This all happens through a manually implemented vtable, which functions similarly to the dynamic dispatch we discussed way back in Chapter 2.

It's a somewhat involved process to construct a Waker, and the mechanics of it aren't all that important for using one, but you can see the building blocks in the RawWakerVTable type in the standard library. It has a constructor that takes the function pointers for wake and wake_by_ref as well as Clone and Drop. The RawWakerVTable, which is usually shared among all of an executor's wakers, is bundled up with a raw pointer intended to hold data specific to each Waker instance (like which future it's for) and is turned into a RawWaker. That is in turn passed to Waker::from_raw to produce a safe Waker that can be passed to Future::poll.

Fulfilling the Poll Contract

So far we've skirted around what a future actually does with a Waker. The idea is fairly simple: if Future::poll returns Poll::Pending, it is the future's responsibility to ensure that *something* calls wake on the provided Waker when the future is next able to make progress. Most futures uphold this property by returning Poll::Pending only if some other future also returned Poll::Pending; in this way, it trivially fulfills the contract of poll since the inner future must follow that same contract. But there can't be turtles all the way down. At some point, you reach a future that does not poll other futures but instead does something like write to a network socket or attempt to receive on a channel. These are commonly referred to as *leaf futures* since they have no children. A leaf future has no inner future but instead directly represents some resource that may not yet be ready to return a result.

NOTE *The poll contract is the reason why the recursive poll call* ❻ *back in Listing 8-4 is necessary for correctness.*

Leaf futures typically come in one of two shapes: those that wait for events that originate within the same process (like a channel receiver), and those that wait for events external to the process (like a TCP packet read). Those that wait for internal events all tend to follow the same pattern: store the Waker where the code that will wake you up can find it, and have that code call wake on the Waker when it generates the relevant event. For example, consider a leaf future that has to wait for a message on an in-memory channel. It stores its Waker inside the part of the channel that is shared between the sender and the receiver and then returns Poll::Pending. When a sender later comes along and injects a message into the channel, it notices the Waker left there by the waiting receiver and calls wake on the Waker before returning from send. Now the receiver is awoken, and the poll contract is upheld.

Leaf futures that deal with external events are more involved, as the code that generates the event they're waiting for knows nothing of futures or wakers. Most often the generating code is the operating system kernel, which knows when a disk is ready or a timer expires, but it could also be a C library that invokes a callback into Rust when an operation completes or some other such external entity. A leaf future wrapping an external resource like this could spin up a thread that executes a blocking system call (or waits for the C callback) and then use the internal waking mechanism, but that would be wasteful; you would spin up a thread every time an operation had to wait and be left with lots of single-use threads sitting around waiting for things.

Instead, executors tend to provide implementations of leaf futures that communicate behind the scenes with the executor to arrange for the appropriate interaction with the operating system. How exactly this is orchestrated depends on the executor and the operating system, but roughly speaking the executor keeps track of all the event sources that it should listen for the next time it goes to sleep. When a leaf future realizes it must wait for an external event, it updates that executor's state (which it knows about since it's provided by the executor crate) to include that external event source alongside its Waker. When the executor can no longer make progress, it gathers all of the event sources the various pending leaf futures are waiting for and does a big blocking call to the operating system, telling it to return when *any* of the resources the leaf futures are waiting on have a new event. On Linux, this is usually achieved with the epoll system call; Windows, the BSDs, macOS, and pretty much every other operating system provide similar mechanisms. When that call returns, the executor calls wake on all the wakers associated with event sources that the operating system reported events for, and thus the poll contract is fulfilled.

NOTE *A* reactor *is the part of an executor that leaf futures register event sources with and that the executor waits on when it has no more useful work to do. It is possible to separate the executor and the reactor, though bundling them together often improves performance as the two can be co-optimized more readily.*

A knock-on effect of the tight integration between leaf futures and the executor is that leaf futures from one executor crate often cannot be used with a different executor. Or at least, they cannot be used unless the leaf future's executor is *also* running. When the leaf future goes to store its Waker and register the event source it's waiting for, the executor it was built against needs to have that state set up and needs to be running so that the event source will actually be monitored and wake eventually called. There are ways around this, such as having leaf futures spawn an executor if one is not already running, but this is not always advisable as it means that an application can transparently end up with multiple executors running at the same time, which can reduce performance and mean you must inspect the state of multiple executors when debugging.

Library crates that wish to support multiple executors have to be generic over their leaf resources. For example, instead of using a particular executor's

TcpStream or File future type, a library can store a generic T: AsyncRead + AsyncWrite. However, the ecosystem has yet to settle on exactly what these traits should look like and which traits are needed, so for the moment it's fairly difficult to make code truly generic over the executor. For example, while AsyncRead and AsyncWrite are somewhat common across the ecosystem (or can be easily adapted if necessary), no traits currently exist for running a future in the background (*spawning*, which we'll discuss later) or for representing a timer.

Waking Is a Misnomer

You may already have realized that Waker::wake doesn't necessarily seem to *wake* anything. For example, for external events (as described in the previous section), the executor is already awake, and it might seem silly for it to then call wake on a Waker that belongs to that executor anyway! The reality is that Waker::wake is a bit of a misnomer—in reality, it signals that a particular future is *runnable*. That is, it tells the executor that it should make sure to poll this particular future when it gets around to it rather than go to sleep again, since this future can make progress. This might wake the executor if it is currently sleeping so it will go poll that future, but that's more of a side effect than its primary purpose.

It is important for the executor to know which futures are runnable for two reasons. First, it needs to know when it can stop polling a future and go to sleep; it's not sufficient to just poll each future until it returns Poll::Pending, since polling a later future might make it possible to progress an earlier future. Consider the case where two futures bounce messages back and forth on channels to one another. When you poll one, the other becomes ready, and vice versa. In this case, the executor should never go to sleep, as there is always more work to do.

Second, knowing which futures are runnable lets the executor avoid polling futures unnecessarily. If an executor manages thousands of pending futures, it shouldn't poll all of them just because an event made one of them runnable. If it did, executing asynchronous code would get very slow indeed.

Tasks and Subexecutors

The futures in an asynchronous program form a tree: a future may contain any number of other futures, which in turn may contain other futures, all the way down to the leaf futures that interact with wakers. The root of each tree is the future you give to whatever the executor's main "run" function is. These root futures are called *tasks*, and they are the only point of contact between the executor and the futures tree. The executor calls poll on the task, and from that point forward the code of each contained future must figure out which inner future(s) to poll in response, all the way down to the relevant leaf.

Executors generally construct a separate Waker for each task they poll so that when wake is later called, they know which task was just made runnable and can mark it as such. That is what the raw pointer in RawWaker is for—to differentiate between tasks while sharing the code for the various Waker methods.

When the executor eventually polls a task, that task starts running from the top of its implementation of Future::poll and must decide from there how to get to the future deeper down that can now make progress. Since each future knows only about its own fields, and nothing about the whole tree, this all happens through calls to poll that each traverse one edge in the tree.

The choice of which inner future to poll is often obvious, but not always. In the case of async/await, the future to poll is the one we're blocked waiting for. But in a future that waits for the first of several futures to make progress (often called a *select*), or for all of a set of futures (often called a *join*), there are many options. A future that has to make such a choice is basically a subexecutor. It could poll all of its inner futures, but doing so could be quite wasteful. Instead, these subexecutors often wrap the Waker they receive in poll's Context with their own Waker type before they invoke poll on any inner future. In the wrapping code, they mark the future they just polled as runnable in their own state before they call wake on the original Waker. That way, when the executor eventually polls the subexecutor future again, the subexecutor can consult its own internal state to figure out which of its inner futures caused the current call to poll, and then only poll those.

BLOCKING IN ASYNC CODE

You must be careful about calling synchronous code from asynchronous code, since any time an executor thread spends executing the current task is time it's *not* spending running other tasks. If a task occupies the current thread for a prolonged period of time without yielding back to the executor, which might happen when executing a blocking system call (like std::sync::sleep), running a subexecutor that doesn't yield occasionally, or running in a tight loop with no awaits, then other tasks the current executor thread is responsible for won't get to run during that time. Usually, this manifests as long delays between when certain tasks can make progress (such as when a client connects) and when they actually get to execute.

Some multithreaded executors implement *work-stealing* techniques, where idle executor threads steal tasks from busy executor threads, but this is more of a mitigation than a solution. Ultimately, you could end up in a situation where all the executor threads are blocked, and thus no tasks get run until one of the blocking operations completes.

In general, you should be very careful with executing compute-intensive operations or calling functions that could block in an asynchronous context. Such operations should either be converted to asynchronous operations where possible or executed on dedicated threads that then communicate using a primitive that does support asynchrony, like a channel. Some executors also provide mechanisms for indicating that a particular segment of asynchronous code might block or for yielding voluntarily in the context of loops that might otherwise not yield, which can compose part of the solution. A good rule of thumb is that no future should be able to run for more than 1 ms without returning Poll::Pending.

Tying It All Together with spawn

When working with asynchronous executors, you may come across an operation that spawns a future. We're now in a position to explore what that means! Let's do so by way of example. First, consider the simple server implementation in Listing 8-14.

```
async fn handle_client(socket: TcpStream) -> Result<()> {
    // Interact with the client over the given socket.
}

async fn server(socket: TcpListener) -> Result<()> {
    while let Some(stream) = socket.accept().await? {
        handle_client(stream).await?;
    }
}
```

Listing 8-14: Handling connections sequentially

The top-level server function is essentially one big future that listens for new connections and does something when a new connection arrives. You hand that future to the executor and say "run this," and since you don't want your program to then exit immediately, you'll probably have the executor block on that future. That is, the call to the executor to run the server future will not return until the server future resolves, which may be never (another client could always arrive later).

Now, every time a new client connection comes in, the code in Listing 8-14 makes a new future (by calling handle_client) to handle that connection. Since the handling is itself a future, we await it and then move on to the next client connection.

The downside of this approach is that we only ever handle one connection at a time—there is no concurrency. Once the server accepts a connection, the handle_client function is called, and since we await it, we don't go around the loop again until handle_client's return future resolves (presumably when that client has left).

We could improve on this by keeping a set of all the client futures and having the loop in which the server accepts new connections also check all the client futures to see if any can make progress. Listing 8-15 shows what that might look like.

```
async fn server(socket: TcpListener) -> Result<()> {
    let mut clients = Vec::new();
    loop {
        poll_client_futures(&mut clients)?;
        if let Some(stream) = socket.try_accept()? {
            clients.push(handle_client(stream));
        }
    }
}
```

Listing 8-15: Handling connections with a manual executor

This at least handles many connections concurrently, but it's quite convoluted. It's also not very efficient because the code now busy-loops, switching between handling the connections we already have and accepting new ones. And it has to check each connection each time, since it won't know which ones can make progress (if any). It also can't await at any point, since that would prevent the other futures from making progress. You could implement your own wakers to ensure that the code polls only the futures that can make progress, but ultimately this is going down the path of developing your own mini-executor.

Another downside of sticking with just the one task for the server that internally contains the futures for all of the client connections is that the server ends up being single-threaded. There is just the one task and to poll it the code must hold an exclusive reference to the task's future (poll takes Pin<&mut Self>), which only one thread can hold at a time.

The solution is to make each client future its own task and leave it to the executor to multiplex among all the tasks. Which, you guessed it, you do by spawning the future. The executor will continue to block on the server future, but if it cannot make progress on that future, it will use its execution machinery to make progress on the other tasks in the meantime behind the scenes. And best of all, if the executor is multithreaded and your client futures are Send, it can run them in parallel since it can hold &muts to the separate tasks concurrently. Listing 8-16 gives an example of what this might look like.

```
async fn server(socket: TcpListener) -> Result<()> {
    while let Some(stream) = socket.accept().await? {
        // Spawn a new task with the Future that represents this client.
        // The current task will continue to just poll for more connections
        // and will run concurrently (and possibly in parallel) with handle_client.
        spawn(handle_client(stream));
    }
}
```

Listing 8-16: Spawning futures to create more tasks that can be polled concurrently

When you spawn a future and thus make it a task, it's sort of like spawning a thread. The future continues running in the background and is multiplexed concurrently with any other tasks given to the executor. However, unlike a spawned thread, spawned tasks still depend on being polled by the executor. If the executor stops running, either because you drop it or because your code no longer runs the executor's code, those spawned tasks will stop making progress. In the server example, imagine what will happen if the main server future resolves for some reason. Since the executor has returned control back to your code, it cannot continue doing, well, anything. Multi-threaded executors often spawn background threads that continue to poll tasks even if the executor yields control back to the user's code, but not all executors do this, so check your executor before you rely on that behavior!

Summary

In this chapter, we've taken a look behind the scenes of the asynchronous constructs available in Rust. We've seen how the compiler implements generators and self-referential types, and why that work was necessary to support what we now know as `async`/`await`. We've also explored how futures are executed, and how wakers allow executors to multiplex among tasks when only some of them can make progress at any given moment. In the next chapter, we'll tackle what is perhaps the deepest and most discussed area of Rust: unsafe code. Take a deep breath, and then turn the page.

9

UNSAFE CODE

The mere mention of unsafe code often elicits strong responses from many in the Rust community, and from many of those watching Rust from the sidelines. While some maintain it's "no big deal," others decry it as "the reason all of Rust's promises are a lie." In this chapter, I hope to pull back the curtain a bit to explain what unsafe is, what it isn't, and how you should go about using it safely. At the time of writing, and likely also when you read this, Rust's precise requirements for unsafe code are still being determined, and even if they were all nailed down, the complete description would be beyond the scope of this book. Instead, I'll do my best to arm you with the building blocks, intuition, and tooling you'll need to navigate your way through most unsafe code.

Your main takeaway from this chapter should be this: unsafe code is the mechanism Rust gives developers for taking advantage of invariants that, for whatever reason, the compiler cannot check. We'll look at the ways in which unsafe does that, what those invariants may be, and what we can do with it as a result.

Throughout this chapter, I'll be talking a lot about invariants. *Invariant* is just a fancy way of saying "something that must be true for your program to be correct." For example, in Rust, one invariant is that references, using & and &mut, do not dangle—they always point to valid values. You can also have application- or library-specific invariants, like "the head pointer is always ahead of the tail pointer" or "the capacity is always a power of two." Ultimately, invariants represent all the assumptions required for your code to be correct. However, you may not always be aware of all the invariants that your code uses, and that's where bugs creep in.

Crucially, unsafe code is not a way to skirt the various rules of Rust, like borrow checking, but rather a way to enforce those rules using reasoning that is beyond the compiler. When you write unsafe code, the onus is on you to ensure that the resulting code is safe. In a way, unsafe is misleading as a keyword when it is used to allow unsafe operations through unsafe {}; it's not that the contained code *is* unsafe, it's that the code is allowed to perform otherwise unsafe operations because in this particular context, those operations *are* safe.

The rest of this chapter is split into four sections. We'll start with a brief examination of how the keyword itself is used, then explore what unsafe allows you to do. Next, we'll look at the rules you must follow in order to write safe unsafe code. Finally, I'll give you some advice about how to actually go about writing unsafe code safely.

The unsafe Keyword

Before we discuss the powers that unsafe grants you, we need to talk about its two different meanings. The unsafe keyword serves a dual purpose in Rust: it marks a particular function as unsafe to call *and* it enables you to invoke unsafe functionality in a particular code block. For example, the method in Listing 9-1 is marked as unsafe, even though it contains no unsafe code. Here, the unsafe keyword serves as a warning to the caller that there are additional guarantees that someone who writes code that invokes decr must manually check.

```
impl<T> SomeType<T> {
    pub unsafe fn decr(&self) {
        self.some_usize -= 1;
    }
}
```

Listing 9-1: An unsafe method that contains only safe code

Listing 9-2 illustrates the second usage. Here, the method itself is not marked as unsafe, even though it contains unsafe code.

```
impl<T> SomeType<T> {
    pub fn as_ref(&self) -> &T {
        unsafe { &*self.ptr }
    }
}
```

Listing 9-2: A safe method that contains unsafe code

These two listings differ in their use of unsafe because they embody different contracts. decr requires the caller to be careful when they call the method, whereas as_ref assumes that the caller *was* careful when invoking other unsafe methods (like decr). To see why, imagine that SomeType is really a reference-counted type like Rc. Even though decr only decrements a number, that decrement may in turn trigger undefined behavior through the safe method as_ref. If you call decr and then drop the second-to-last Rc of a given T, the reference count drops to zero and the T will be dropped—but the program might still call as_ref on the last Rc, and end up with a dangling reference.

NOTE *Undefined behavior describes the consequences of a program that violates invariants of the language at runtime. In general, if a program triggers undefined behavior, the outcome is entirely unpredictable. We'll cover undefined behavior in greater detail later in this chapter.*

Conversely, as long as there is no way to corrupt the Rc reference count using safe code, it is always safe to dereference the pointer inside the Rc the way the code for as_ref does—the fact that &self exists is proof that the pointer must still be valid. We can use this to give the caller a safe API to an otherwise unsafe operation, which is a core piece of how to use unsafe responsibly.

For historical reasons, every unsafe fn contains an implicit unsafe block in Rust today. That is, if you declare an unsafe fn, you can always invoke any unsafe methods or primitive operations inside that fn. However, that decision is now considered a mistake, and it's currently being reverted through the already accepted and implemented RFC 2585. This RFC warns about having an unsafe fn that performs unsafe operations without an explicit unsafe block inside it. The lint will also likely become a hard error in future editions of Rust. The idea is to reduce the "footgun radius"—if every unsafe fn is one giant unsafe block, then you might accidentally perform unsafe operations without realizing it! For example, in decr in Listing 9-1, under the current rules you could also have added *std::ptr::null() without any unsafe annotation.

The distinction between unsafe as a marker and unsafe blocks as a mechanism to enable unsafe operations is important, because you must think about them differently. An unsafe fn indicates to the caller that they have to be careful when calling the fn in question and that they must ensure that the function's documented safety invariants hold.

Meanwhile, an unsafe block implies that whoever wrote that block carefully checked that the safety invariants for any unsafe operations performed inside it hold. If you want an approximate real-world analogy, unsafe fn is an unsigned contract that asks the author of calling code to "solemnly swear X, Y, and Z." Meanwhile, unsafe {} is the calling code's author signing off on all the unsafe contracts contained within the block. Keep that in mind as we go through the rest of this chapter.

Great Power

So, once you sign the unsafe contract with unsafe {}, what are you allowed to do? Honestly, not that much. Or rather, it doesn't enable that many new features. Inside an unsafe block, you are allowed to dereference raw pointers and call unsafe fns.

That's it. Technically, there are a few other things you can do, like accessing mutable and external static variables and accessing fields of unions, but those don't change the discussion much. And honestly, that's enough. Together, these powers allow you to wreak all sorts of havoc, like turning types into one another with mem::transmute, dereferencing raw pointers that point to who knows where, casting &'a to &'static, or making types shareable across thread boundaries even though they're not thread-safe.

In this section, we won't worry too much about what can go wrong with these powers. We'll leave that for the boring, responsible, grown-up section that comes after. Instead, we'll look at these neat shiny new toys and what we can do with them.

Juggling Raw Pointers

One of the most fundamental reasons to use unsafe is to deal with Rust's raw pointer types: *const T and *mut T. You should think of these as more or less analogous to &T and &mut T, except that they don't have lifetimes and are not subject to the same validity rules as their & counterparts, which we'll discuss later in the chapter. These types are interchangeably referred to as *pointers* and *raw pointers*, mostly because many developers instinctively refer to references as pointers, and calling them raw pointers makes the distinction clearer.

Since fewer rules apply to * than &, you can cast a reference to a pointer even outside an unsafe block. Only if you want to go the other way, from * to &, do you need unsafe. You'll generally turn a pointer back into a reference to do useful things with the pointed-to data, such as reading or modifying its value. For that reason, a common operation to use on pointers is unsafe { &*ptr } (or &mut *). The * there may look strange as the code is just constructing a reference, not dereferencing the pointer, but it makes sense if you look at the types; if you have a *mut T and want a &mut T, then &mut ptr would just give you a &mut *mut T. You need the * to indicate that you want the mutable reference to what ptr is a pointer *to*.

Unrepresentable Lifetimes

As raw pointers do not have lifetimes, they can be used in circumstances where the liveness of the value being pointed to cannot be expressed statically within Rust's lifetime system, such as a self-pointer in a self-referential struct like the generators we discussed in Chapter 8. A pointer that points into self is valid for as long as self is around (and doesn't move, which is what Pin is for), but that isn't a lifetime you can generally name. And while the entire self-referential type may be 'static, the self-pointer isn't—if it were static, then even if you gave away that pointer to someone else, they could continue to use it forever, even after self was gone! Take the type in Listing 9-3 as an example; here we attempt to store the raw bytes that make up a value alongside its stored representation.

```
struct Person<'a> {
    name: &'a str,
    age: usize,
}
struct Parsed {
    bytes: [u8; 1024],
    parsed: Person<'???>,
}
```

Listing 9-3: Trying, and failing, to name the lifetime of a self-referential reference

The reference inside `Person` wants to refer to data stored in `bytes` in `Parsed`, but there is no lifetime we can assign to that reference from `Parsed`. It's not `'static` or something like `'self` (which doesn't exist), because if `Parsed` is moved, the reference is no longer valid.

Since pointers do not have lifetimes, they circumvent this problem because you don't have to be able to name the lifetime. Instead, you just have to make sure that when you do use the pointer, it's still valid, which is what you sign off on when you write `unsafe { &*ptr }`. In the example in Listing 9-3, `Person` would instead store a `*const str` and then unsafely turn that into a `&str` at the appropriate times when it can guarantee that the pointer is still valid.

A similar issue arises with a type like `Arc`, which has a pointer to a value that's shared for some duration, but that duration is known only at runtime when the last `Arc` is dropped. The pointer is kind-of, sort-of `'static`, but not really—like in the self-referential case, the pointer is no longer valid when the last `Arc` reference goes away, so the lifetime is more like `'self`. In `Arc`'s cousin, `Weak`, the lifetime is also "when the last `Arc` goes away," but since a `Weak` isn't an `Arc`, the lifetime isn't even tied to `self`. So, `Arc` and `Weak` both use raw pointers internally.

Pointer Arithmetic

With raw pointers, you can do arbitrary pointer arithmetic, just like you can in C, by using `.offset()`, `.add()`, and `.sub()` to move the pointer to any byte that lives within the same allocation. This is most often used in highly space-optimized data structures, like hash tables, where storing an extra pointer for each element would add too much overhead and using slices isn't possible. Those are fairly niche use cases, and we won't be talking more about them in this book, but I encourage you to read the code for `hashbrown::RawTable` (*https://github.com/rust-lang/hashbrown/*) if you want to learn more!

The pointer arithmetic methods are unsafe to call even if you don't want to turn the pointer into a reference afterwards. There are a couple of reasons for this, but the main one is that it is illegal to make a pointer point beyond the end of the allocation that it originally pointed to. Doing so triggers undefined behavior, and the compiler is allowed to decide to eat your code and replace it with arbitrary nonsense that only a compiler could understand. If you do use these methods, read the documentation carefully!

To Pointer and Back Again

Often when you need to use pointers, it's because you have some normal Rust type, like a reference, a slice, or a string, and you have to move to the world of pointers for a bit and then go back to the original normal type. Some of the key standard library types therefore provide you with a way to turn them into their raw constituent parts, such as a pointer and a length for a slice, and a way to turn them back into the whole using those same parts. For example, you can get a slice's data pointer with `as_ptr` and its length with `[]::len`. You can then reconstruct the slice by providing those same values to `std::slice::from_raw_parts`. `Vec`, `Arc`, and `String` have similar methods that return a raw pointer to the underlying allocation, and `Box` has `Box::into_raw` and `Box::from_raw`, which do the same thing.

Playing Fast and Loose with Types

Sometimes, you have a type `T` and want to treat it as some other type `U`. Whether that's because you need to do lightning-fast zero-copy parsing or because you need to fiddle with some lifetimes, Rust provides you with some (very unsafe) tools to do so.

The first and by far most widely used of these is pointer casting: you can cast a `*const T` to any other `*const U` (and the same for `mut`), and you don't even need `unsafe` to do it. The unsafety comes into play only when you later try to use the cast pointer as a reference, as you have to assert that the raw pointer can in fact be used as a reference to the type it's pointing to.

This kind of pointer type casting comes in particularly handy when working with foreign function interfaces (FFI)—you can cast any Rust pointer to a `*const std::ffi::c_void` or `*mut std::ffi::c_void`, and then pass that to a C function that expects a void pointer. Similarly, if you get a void pointer from C that you previously passed in, you can trivially cast it back into its original type.

Pointer casts are also useful when you want to interpret a sequence of bytes as plain old data—types like integers, Booleans, characters, and arrays, or `#[repr(C)]` structs of these—or write such types directly out as a byte stream without serialization. There are a lot of safety invariants to keep in mind if you want to try to do that, but we'll leave that for later.

Calling Unsafe Functions

Arguably `unsafe`'s most commonly used feature is that it enables you to call unsafe functions. Deeper down the stack, most of those functions are unsafe because they operate on raw pointers at some fundamental level, but higher up the stack you tend to interact with unsafety primarily through function calls.

There's really no limit to what calling an unsafe function might enable, as it is entirely up to the libraries you interact with. But *in general,* unsafe functions can be divided into three camps: those that interact with non-Rust interfaces, those that skip safety checks, and those that have custom invariants.

Foreign Function Interfaces

Rust lets you declare functions and static variables that are defined in a language other than Rust using `extern` blocks (which we'll discuss at length in Chapter 11). When you declare such a block, you're telling Rust that the items appearing within it will be implemented by some external source when the final program binary is linked, such as a C library you are integrating with. Since externs exist outside of Rust's control, they are inherently unsafe to access. If you call a C function from Rust, all bets are off—it might overwrite your entire memory contents and clobber all your neatly arranged references into random pointers into the kernel somewhere. Similarly, an extern static variable could be modified by external code at any time, and could be filled with all sorts of bad bytes that don't reflect its declared type at all. In an unsafe block, though, you can access externs to your heart's delight, as long as you're willing to vouch for the other side of the extern behaving according to Rust's rules.

I'll Pass on Safety Checks

Some unsafe operations can be made entirely safe by introducing additional runtime checks. For example, accessing an item in a slice is unsafe since you might try to access an item beyond the length of the slice. But, given how common the operation is, it'd be unfortunate if indexing into a slice was unsafe. Instead, the safe implementation includes bounds checks that (depending on the method you use) either panic or return an Option if the index you provide is out of bounds. That way, there is no way to cause undefined behavior even if you pass in an index beyond the slice's length. Another example is in hash tables, which hash the key you provide rather than letting you provide the hash yourself; this ensures that you'll never try to access a key using the wrong hash.

However, in the endless pursuit of ultimate performance, some developers may find these safety checks add just a little too much overhead in their tightest loops. To cater to situations where peak performance is paramount and the caller knows that the indexes are in bounds, many data structures provide alternate versions of particular methods without these safety checks. Such methods usually include the word unchecked in the name to indicate that they blindly trust the provided arguments to be safe and that they do not do any of those pesky, slow safety checks. Some examples are NonNull::new_unchecked, slice::get_unchecked, NonZero::new_unchecked, Arc::get _mut_unchecked, and str::from_utf8_unchecked.

In practice, the safety and performance trade-off for unchecked methods is rarely worth it. As always with performance optimization, measure first, then optimize.

Custom Invariants

Most uses of unsafe rely on custom invariants to some degree. That is, they rely on invariants beyond those provided by Rust itself, which are specific to the particular application or library. Since so many functions fall into this category, it's hard to give a good general summary of this class of unsafe functions. Instead, I'll give some examples of unsafe functions with custom invariants that you may come across in practice and want to use:

MaybeUninit::assume_init

The MaybeUninit type is one of the few ways in which you can store values that are not valid for their type in Rust. You can think of a MaybeUninit<T> as a T that may not be legal to use as a T at the moment. For example, a MaybeUninit<NonNull> is allowed to hold a null pointer, a MaybeUninit<Box> is allowed to hold a dangling heap pointer, and a

`MaybeUninit<bool>` is allowed to hold the bit pattern for the number 3 (normally it must be 0 or 1). This comes in handy if you are constructing a value bit by bit or are dealing with zeroed or uninitialized memory that will eventually be made valid (such as by being filled through a call to `std::io::Read::read`). The `assume_init` function asserts that the `MaybeUninit` now holds a valid value for the type `T` and can therefore be used as a `T`.

ManuallyDrop::drop

The `ManuallyDrop` type is a wrapper type around a type `T` that does not drop that `T` when the `ManuallyDrop` is dropped. Or, phrased differently, it decouples the dropping of the outer type (`ManuallyDrop`) from the dropping of the inner type (`T`). It implements safe access to the `T` through `DerefMut<Target = T>` but also provides a `drop` method (separately from the `drop` method of the `Drop` trait) to drop the wrapped `T` *without* dropping the `ManuallyDrop`. That is, the `drop` function takes `&mut self` despite dropping the `T`, and so leaves the `ManuallyDrop` behind. This comes in handy if you have to explicitly drop a value that you cannot move, such as in implementations of the `Drop` trait. Once that value is dropped, it is no longer safe to try to access the `T`, which is why the call to `drop` is unsafe—it asserts that the `T` will never be accessed again.

std::ptr::drop_in_place

`drop_in_place` lets you call a value's destructor directly through a pointer to that value. This is unsafe because the pointee will be left behind after the call, so if some code then tries to dereference the pointer, it'll be in for a bad time! This method is particularly useful when you may want to reuse memory, such as in an arena allocator, and need to drop an old value in place without reclaiming the surrounding memory.

Waker::from_raw

In Chapter 8 we talked about the `Waker` type and how it is made up of a data pointer and a `RawWaker` that holds a manually implemented vtable. Once a `Waker` has been constructed, the raw function pointers in the vtable, such as `wake` and `drop`, can be called from safe code (through `Waker::wake` and `drop(waker)`, respectively). `Waker::from_raw` is where the asynchronous executor asserts that all the pointers in its vtable are in fact valid function pointers that follow the contract set forth in the documentation of `RawWakerVTable`.

std::hint::unreachable_unchecked

The `hint` module holds functions that give hints to the compiler about the surrounding code but do not actually produce any machine code. The `unreachable_unchecked` function in particular tells the compiler that it is impossible for the program to reach a section of the code at runtime. This in turn allows the compiler to make optimizations based on that knowledge, such as eliminating conditional branches to that location. Unlike the unreachable! macro, which panics if the code does reach the line in question, the effects of an erroneous unreachable_unchecked are hard to predict. The compiler optimizations may cause peculiar and hard-to-debug behavior, not to mention that your program will continue running when something it believed to be true was not!

std::ptr::{read,write}_{unaligned,volatile}

The ptr module holds a number of functions that let you work with *odd* pointers—those that do not meet the assumptions that Rust generally makes about pointers. The first of these functions are read_unaligned and write_unaligned, which let you access pointers that point to a T even if that T is not stored according to T's alignment (see the section on alignment in Chapter 2). This might happen if the T is contained directly in a byte array or is otherwise packed in with other values without proper padding. The second notable pair of functions is read_volatile and write_volatile, which let you operate on pointers that don't point to normal memory. Concretely, these functions will always access the given pointer (they won't be cached in a register, for example, even if you read the same pointer twice in a row), and the compiler won't reorder the volatile accesses relative to other volatile accesses. Volatile operations come in handy when working with pointers that aren't backed by normal DRAM memory—we'll discuss this further in Chapter 11. Ultimately, these methods are unsafe because they dereference the given pointer (and to an owned T, at that), so you as the caller need to sign off on all the contracts associated with doing so.

std::thread::Builder::spawn_unchecked

The normal thread::spawn that we know and love requires that the provided closure is 'static. That bound stems from the fact that the spawned thread might run for an indeterminate amount of time; if we were allowed to use a reference to, say, the caller's stack, the caller might return well before the spawned thread exits, rendering the reference invalid. Sometimes, however, you know that some non-'static value in the caller will outlive the spawned thread. This might happen if you join the thread before dropping the value in question, or if the value is dropped only strictly after you know the spawned thread will no longer use it. That's where spawn_unchecked comes in—it does not have the 'static bound and thus lets you implement those use cases as long as you're willing to sign the contract saying that no unsafe accesses will happen as a result. Be careful of panics, though; if the caller panics, it might drop values earlier than you planned and cause undefined behavior in the spawned thread!

Note that all of these methods (and indeed all unsafe methods in the standard library) provide explicit documentation for their safety invariants, as should be the case for any unsafe method.

Implementing Unsafe Traits

Unsafe traits aren't unsafe to *use*, but unsafe to *implement*. This is because unsafe code is allowed to rely on the correctness (defined by the trait's documentation) of the implementation of unsafe traits. For example, to implement the unsafe trait Send, you need to write `unsafe impl Send for ....` Like unsafe functions, unsafe traits generally have custom invariants that are (or at least should be) specified in the documentation for the trait. Thus, it's difficult to cover unsafe traits as a group, so here too I'll give some common examples from the standard library that are worth going over.

Send and Sync

The Send and Sync traits denote that a type is safe to send or share across thread boundaries, respectively. We'll talk more about these traits in Chapter 10, but for now what you need to know is that they are auto-traits, and so they'll usually be implemented for most types for you by the compiler. But, as tends to be the case with auto-traits, Send and Sync will not be implemented if any members of the type in question are not themselves Send or Sync.

In the context of unsafe code, this problem occurs primarily due to raw pointers, which are neither Send nor Sync. At first glance, this might seem reasonable: the compiler has no way to know who else may have a raw pointer to the same value or how they may be using it at the moment, so how can the type be safe to send across threads? Now that we're seasoned unsafe developers though, that argument seems weak—after all, dereferencing a raw pointer is already unsafe, so why should handling the invariants of Send and Sync be any different?

Strictly speaking, raw pointers could be both Send and Sync. The problem is that if they were, the types that contain raw pointers would automatically be Send and Sync themselves, even though their author might not realize that was the case. The developer might then unsafely dereference the raw pointers without ever thinking about what would happen if those types were sent or shared across thread boundaries, and thus inadvertently introduce undefined behavior. Instead, the raw pointer types block these automatic implementations as an additional safeguard to unsafe code to make authors explicitly sign the contract that they have also followed the Send and Sync invariants.

> **NOTE** *A common mistake with unsafe implementations of Send and Sync is to forget to add bounds to generic parameters: unsafe impl<T: Send> Send for MyUnsafeType<T> {}.*

GlobalAlloc

The GlobalAlloc trait is how you implement a custom memory allocator in Rust. We won't talk too much about that topic in this book, but the trait itself is interesting. Listing 9-4 gives the required methods for the GlobalAlloc trait.

```
pub unsafe trait GlobalAlloc {
    pub unsafe fn alloc(&self, layout: Layout) -> *mut u8;
    pub unsafe fn dealloc(&self, ptr: *mut u8, layout: Layout);
}
```

Listing 9-4: The `GlobalAlloc` trait with its required methods

At its core, the trait has one method for allocating a new chunk of memory, `alloc`, and one for deallocating a chunk of memory, `dealloc`. The `Layout` argument describes the type's size and alignment, as we discussed in Chapter 2. Each of those methods is unsafe and carries a number of safety invariants that its callers must uphold.

`GlobalAlloc` itself is also unsafe because it places restrictions on the implementer of the trait, not the caller of its methods. Only the unsafety of the trait ensures that implementers agree to uphold the invariants that Rust itself assumes of its memory allocator, such as in the standard library's implementation of `Box`. If the trait was not unsafe, an implementer could safely implement `GlobalAlloc` in a way that produced unaligned pointers or incorrectly sized allocations, which would trigger unsafety in otherwise safe code that assumes that allocations are sane. This would break the rule that safe code should not be able to trigger memory unsafety in other safe code, and thus cause all sorts of mayhem.

Surprisingly Not Unpin

The `Unpin` trait is not unsafe, which comes as a surprise to many Rust developers. It may even come as a surprise to you after reading Chapter 8. After all, the trait is supposed to ensure that self-referential types aren't invalidated if they're moved after they have established internal pointers (that is, after they've been placed in a `Pin`). It seems strange, then, that `Unpin` can be used to safely remove a type from a `Pin`.

There are two main reasons why `Unpin` isn't an unsafe trait. First, it's unnecessary. Implementing `Unpin` for a type that you control does not grant you the ability to safely pin or unpin a `!Unpin` type; that still requires unsafety in the form of a call to `Pin::new_unchecked` or `Pin::get_unchecked_mut`. Second, there is already a safe way for you to unpin any type you control: the `Drop` trait! When you implement `Drop` for a type, you're passed `&mut self`, even if your type was previously stored in a `Pin` and is `!Unpin`, all without any unsafety. That potential for unsafety is covered by the invariants of `Pin::new_unchecked`, which must be upheld to create a `Pin` of such an `!Unpin` type in the first place.

When to Make a Trait Unsafe

Few traits in the wild are unsafe, but those that are all follow the same pattern. A trait should be unsafe if safe code that assumes that trait is implemented correctly can exhibit memory unsafety if the trait is *not* implemented correctly.

The `Send` trait is a good example to keep in mind here—safe code can easily spawn a thread and pass a value to that spawned thread, but if `Rc` were

Send, that sequence of operations could trivially lead to memory unsafety. Consider what would happen if you cloned an Rc<Box> and sent it to another thread: the two threads could easily both try to deallocate the Box since they do not correctly synchronize access to the Rc's reference count.

The Unpin trait is a good counterexample. While it is possible to write unsafe code that triggers memory unsafety if Unpin is implemented incorrectly, no entirely safe code can trigger memory unsafety due to an implementation of Unpin. It's not always easy to determine that a trait can be safe (indeed, the Unpin trait was unsafe throughout most of the RFC process), but you can always err on the side of making the trait unsafe, and then make it safe later on if you realize that is the case! Just keep in mind that that is a backward incompatible change.

Also keep in mind that just because it feels like an incorrect (or even malicious) implementation of a trait would cause a lot of havoc, that's not necessarily a good reason to make it unsafe. The unsafe marker should first and foremost be used to highlight cases of *memory* unsafety, not just something that can trigger errors in business logic. For example, the Eq, Ord, Deref, and Hash traits are all safe, even though there is likely much code out in the world that would go haywire if faced with a malicious implementation of, say, Hash that returned a different random hash each time it was called. This extends to unsafe code too—there is almost certainly unsafe code out there that would be memory-unsafe in the presence of such an implementation of Hash—but that does not mean Hash should be unsafe. The same is true for an implementation of Deref that dereferenced to a different (but valid) target each time. Such unsafe code would be relying on a contract of Hash or Deref that does not actually hold; Hash never claimed that it was deterministic, and neither did Deref. Or rather, the authors of those implementations never used the unsafe keyword to make that claim!

NOTE *An important implication of traits like Eq, Hash, and Deref being safe is that unsafe code can rely only on the safety of safe code, not its correctness. This applies not only to traits, but to all unsafe/safe code interactions.*

Great Responsibility

So far, we've looked mainly at the various things that you are allowed to do with unsafe code. But unsafe code is allowed to do those things only if it does so safely. Even though unsafe code can, say, dereference a raw pointer, it must do so only if it knows that pointer is valid as a reference to its pointee at that moment in time, subject to all of Rust's normal requirements of references. In other words, unsafe code is given access to tools that could be used to do unsafe things, but it must do only safe things using those tools.

That, then, raises the question of what *safe* even means in the first place. When is it safe to dereference a pointer? When is it safe to transmute between two different types? In this section, we'll explore some of the key invariants to keep in mind when wielding the power of unsafe, look at some common gotchas, and get familiar with some of the tools that help you write safer unsafe code.

The exact rules around what it means for Rust code to be safe are still being worked out. At the time of writing, the Unsafe Code Guidelines Working Group is hard at work nailing down all the dos and don'ts, but many questions remain unanswered. Most of the advice in this section is more or less settled, but I'll make sure to call out any that isn't. If anything, I'm hoping that this section will teach you to be careful about making assumptions when you write unsafe code, and prompt you to double-check the Rust reference before you declare your code production-ready.

What Can Go Wrong?

We can't really get into the rules unsafe code must abide by without talking about what happens if you violate those rules. Let's say you do mutably access a value from multiple threads concurrently, construct an unaligned reference, or dereference a dangling pointer—now what?

Unsafe code that is not ultimately safe is referred to as having *undefined behavior*. Undefined behavior generally manifests in one of three ways: not at all, through visible errors, or through invisible corruption. The first is the happy case—you wrote some code that is truly not safe, but the compiler generated sane code that the computer you're running the code on executes in a sane way. Unfortunately, the happiness here is very brittle. Should a new and slightly smarter version of the compiler come along, or some surrounding code cause the compiler to apply another optimization, the code may no longer do something sane and tip over into one of the worse cases. Even if the same code is compiled by the same compiler, if it runs on a different platform or host, the program might act differently! This is why it is important to avoid undefined behavior even if everything currently seems to work fine. Not to do so is like playing a second round of Russian roulette just because you survived the first.

Visible errors are the easiest undefined behavior to catch. If you dereference a null pointer, for example, your program will (in all likelihood) crash with an error, which you can then debug back to the root cause. That debugging may itself be difficult, but at least you have a notification that something is wrong. Visible errors can also manifest in less severe ways, such as deadlocks, garbled output, or panics that are printed but don't trigger a program exit, all of which tell you that there is a bug in your code that you have to go fix.

The worst manifestation of undefined behavior is when there is no immediate visible effect, but the program state is invisibly corrupted. Transaction amounts might be slightly off from what they should be, backups might be silently corrupted, or random bits of internal memory could be exposed to external clients. The undefined behavior could cause ongoing corruption, or extremely infrequent outages. Part of the challenge with undefined behavior is that, as the name implies, the behavior of the nonsafe unsafe code is not defined—the compiler might eliminate it entirely, dramatically change the semantics of the code, or even miscompile surrounding code. What that does to your program is entirely dependent on what the code in question does. The unpredictable impact of undefined behavior is the reason why *all* undefined behavior should be considered a serious bug, no matter how it *currently* manifests.

WHY UNDEFINED BEHAVIOR?

An argument that often comes up in conversations about undefined behavior is that the compiler should emit an error if code exhibits undefined behavior instead of doing something weird and unpredictable. That way, it would be near-impossible to write bad unsafe code!

Unfortunately, that would be impossible because undefined behavior is rarely explicit or obvious. Instead, what usually happens is that the compiler simply applies optimizations under the assumption that the code follows the specification. Should that turn out to not be the case—which is rarely clear until runtime—it's difficult to predict what the effect might be. Maybe the optimization is still valid, and nothing bad happens; but maybe it's not, and the semantics of the code end up slightly different from that of the unoptimized version.

If we were to tell compiler developers that they aren't allowed to assume anything about the underlying code, what we'd really be telling them is that they cannot perform a wide range of the optimizations that they implement with great success today. Nearly all sophisticated optimizations make assumptions about what the code in question can and cannot do according to the language specification.

If you want a good illustration of how specifications and compiler optimizations interact in strange ways where it's hard to assign blame, I recommend reading through Ralf Jung's blog post "We Need Better Language Specs" (*https://www.ralfj.de/blog/2020/12/14/provenance.html*).

Validity

Perhaps the most important concept to understand before writing unsafe code is *validity*, which dictates the rules for what values inhabit a given type—or, less formally, the rules for a type's values. The concept is simpler than it sounds, so let's dive into some concrete examples.

Reference Types

Rust is very strict about what values its reference types can hold. Specifically, references must never dangle, must always be aligned, and must always point to a valid value for their target type. In addition, a shared and an exclusive reference to a given memory location can never exist at the same time, and neither can multiple exclusive references to a location. These rules apply regardless of whether your code uses the references or not—you are not allowed to create a null reference even if you then immediately discard it!

Shared references have the additional constraint that the pointee is not allowed to change during the reference's lifetime. That is, any value the pointee contains must remain exactly the same over its lifetime. This applies transitively, so if you have an & to a type that contains a *mut T, you are not allowed to ever mutate the T through that *mut even though you could write code to do so using unsafe. The *only* exception to this rule is a value wrapped by the UnsafeCell type. All other types that provide interior mutability, like Cell, RefCell, and Mutex, internally use an UnsafeCell.

An interesting result of Rust's strict rules for references is that for many years, it was impossible to safely take a reference to a field of a packed or partially uninitialized struct that used repr(Rust). Since repr(Rust) leaves a type's layout undefined, the only way to get the address of a field was by writing &some_struct.field as *const _. However, if some_struct is packed, then some_struct.field may not be aligned, and thus creating an & to it is illegal! Further, if some_struct isn't fully initialized, then the some_struct reference itself cannot exist! In Rust 1.51.0, the ptr::addr_of! macro was stabilized, which added a mechanism for directly obtaining a reference to a field without first creating a reference, fixing this particular problem. Internally, it is implemented using something called *raw references* (not to be confused with raw pointers), which directly create pointers to their operands rather than going via a reference. Raw references were introduced in RFC 2582 but haven't been stabilized themselves yet at the time of writing.

Primitive Types

Some of Rust's primitive types have restrictions on what values they can hold. For example, a bool is defined as being 1 byte large but is only allowed to hold the value 0x00 or the value 0x01, and a char is not allowed to hold a surrogate or a value above char::MAX. Most of Rust's primitive types, and indeed most of Rust's types overall, also cannot be constructed from uninitialized memory. These restrictions may seem arbitrary, but again often stem from the need to enable optimizations that wouldn't be possible otherwise.

A good illustration of this is the niche optimization, which we discussed briefly when talking about pointer types earlier in this chapter. To recap, the niche optimization tucks away the enum discriminant value in the wrapped type in certain cases. For example, since a reference cannot ever be all zeros, an Option<&T> can use all zeros to represent None, and thus avoid spending an extra byte (plus padding) to store the discriminator byte. The compiler can optimize Booleans in the same way and potentially take it even further. Consider the type Option<Option<bool>>>. Since the compiler knows that the bool is either 0x00 or 0x01, it's free to use 0x02 to represent Some(None) and 0x03 to represent None. Very nice and tidy! But if someone were to come along and treat the byte 0x03 as a bool, and then place that value in an Option<Option<bool>> optimized in this way, bad things would happen.

It bears repeating that it's not important whether the Rust compiler currently implements this optimization or not. The point is that it is allowed to, and therefore any unsafe code you write must conform to that contract or risk hitting a bug later on should the behavior change.

Owned Pointer Types

Types that point to memory they own, like Box and Vec, are generally subject to the same optimizations as if they held an exclusive reference to the pointed-to memory unless they're explicitly accessed through a shared reference. Specifically, the compiler assumes that the pointed-to memory is not shared or aliased elsewhere, and makes optimizations based on that assumption. For example, if you extracted the pointer from a Box and then constructed two Boxes from that same pointer and wrapped them in ManuallyDrop to prevent a double-free, you'd likely be entering undefined behavior territory. That's the case even if you only ever access the inner type through shared references. (I say "likely" because this isn't fully settled in the language reference yet, but a rough consensus has arisen.)

Storing Invalid Values

Sometimes you need to store a value that isn't currently valid for its type. The most common example of this is if you want to allocate a chunk of memory for some type T and then read in the bytes from, say, the network. Until all the bytes have been read in, the memory isn't going to be a valid T. Even if you just tried to read the bytes into a slice of u8, you would have to zero those u8s first, because constructing a u8 from uninitialized memory is also undefined behavior.

The MaybeUninit<T> type is Rust's mechanism for working with values that aren't valid. A MaybeUninit<T> stores exactly a T (it is #[repr(transparent)]), but the compiler knows to make no assumptions about the validity of that T. It won't assume that references are non-null, that a Box<T> isn't dangling, or that a bool is either 0 or 1. This means it's safe to hold a T backed by uninitialized memory inside a MaybeUninit (as the name implies). MaybeUninit is also a very useful tool in other unsafe code where you have to temporarily store a value that may be invalid. Maybe you have to store an aliased Box<T> or stash a char surrogate for a second—MaybeUninit is your friend.

You will generally do only three things with a MaybeUninit: create it using the MaybeUninit::uninit method, write to its contents using MaybeUninit::as _mut_ptr, or take the inner T once it is valid again with MaybeUninit::assume_init. As its name implies, uninit creates a new MaybeUninit<T> of the same size as a T that initially holds uninitialized memory. The as_mut_ptr method gives you a raw pointer to the inner T that you can then write to; nothing stops you from reading from it, but reading from any of the uninitialized bits is undefined behavior. And finally, the unsafe assume_init method consumes the MaybeUninit<T> and returns its contents as a T following the assertion that the backing memory now makes up a valid T.

Listing 9-5 shows an example of how we might use MaybeUninit to safely initialize a byte array without explicitly zeroing it.

```
fn fill(gen: impl FnMut() -> Option<u8>) {
    let mut buf = [MaybeUninit::<u8>::uninit(); 4096];
    let mut last = 0;
    for (i, g) in std::iter::from_fn(gen).take(4096).enumerate() {
        buf[i] = MaybeUninit::new(g);
        last = i + 1;
    }
    // Safety: all the u8s up to last are initialized.
let init: &[u8] = unsafe {
  MaybeUninit::slice_assume_init_ref(&buf[..last])
};
    // ... do something with init ...
}
```

Listing 9-5: Using MaybeUninit to safely initialize an array

While we could have declared buf as [0; 4096] instead, that would require the function to first write out all those zeros to the stack before executing, even if it's going to overwrite them all again shortly thereafter. Normally that wouldn't have a noticeable impact on performance, but if this was in a sufficiently hot loop, it might! Here, we instead allow the array to keep whatever values happened to be on the stack when the function was called, and then overwrite only what we end up needing.

NOTE   *Be careful with dropping partially initialized memory. If a panic causes an unexpected early drop before the MaybeUninit<T> has been fully initialized, you'll have to take care to drop only the parts of T that are now valid, if any. You can just drop the MaybeUninit and have the backing memory forgotten, but if it holds, say, a Box, you might end up with a memory leak!*

Panics

An important and often overlooked aspect of ensuring that code using unsafe operations is safe is that the code must also be prepared to handle panics. In particular, as we discussed briefly in Chapter 5, Rust's default panic handler on most platforms will not crash your program on a panic but will instead *unwind* the current thread. An unwinding panic effectively drops everything in the current scope, returns from the current function, drops everything in the scope that enclosed the function, and so on, all the way down the stack until it hits the first stack frame for the current thread. If you don't take unwinding into account in your unsafe code, you may be in for trouble. For example, consider the code in Listing 9-6, which tries to efficiently push many values into a Vec at once.

```
impl<T: Default> Vec<T> {
    pub fn fill_default(&mut self) {
        let fill = self.capacity() - self.len();
        if fill == 0 { return; }
        let start = self.len();
        unsafe {
            self.set_len(start + n);
```

```
        for i in 0..fill {
            *self.get_unchecked_mut(start + i) = T::default();
        }
    }
  }
}
```

Listing 9-6: A seemingly safe method for filling a vector with Default *values*

Consider what happens to this code if a call to T::default panics. First, fill_default will drop all its local values (which are just integers) and then return. The caller will then do the same. At some point up the stack, we get to the owner of the Vec. When the owner drops the vector, we have a problem: the length of the vector now indicates that we own more Ts than we actually produced due to the call to set_len. For example, if the very first call to T::default panicked when we aimed to fill eight elements, that means Vec::drop will call drop on eight Ts that actually contain uninitialized memory!

The fix in this case is simple: the code must update the length *after* writing all the elements. We wouldn't have realized there was a problem if we didn't carefully consider the effect of unwinding panics on the correctness of our unsafe code.

When you're combing through your code for these kinds of problems, you'll want to look out for any statements that may panic, and consider whether your code is safe if they do. Alternatively, check whether you can convince yourself that the code in question will never panic. Pay particular attention to anything that calls user-provided code—in those cases, you have no control over the panics and should assume that the user code will panic.

A similar situation arises when you use the ? operator to return early from a function. If you do this, make sure that your code is still safe if it does not execute the remainder of the code in the function. It's rarer for ? to catch you off guard since you opted into it explicitly, but it's worth keeping an eye out for.

Casting

As we discussed in Chapter 2, two different types that are both #[repr(Rust)] may be represented differently in memory even if they have fields of the same type and in the same order. This in turn means that it's not always obvious whether it is safe to cast between two different types. In fact, Rust doesn't even guarantee that two instances of a single type with generic arguments that are themselves laid out the same way are represented the same way. For example, in Listing 9-7, A and B are not guaranteed to have the same in-memory representation.

```
struct Foo<T> {
    one: bool,
    two: PhantomData<T>,
}
struct Bar;
struct Baz;
```

```
type A = Foo<Bar>;
type B = Foo<Baz>;
```

Listing 9-7: Type layout is not predictable.

The lack of guarantees for repr(Rust) is important to keep in mind when you do type casting in unsafe code—just because two types feel like they should be interchangeable, that is not necessarily the case. Casting between two types that have different representations is a quick path to undefined behavior. At the time of writing, the Rust community is actively working out the exact rules for how types are represented, but for now, very few guarantees are given, so that's what we have to work with.

Even if identical types were guaranteed to have the same in-memory representation, you'd still run into the same problem when types are nested. For example, while UnsafeCell<T>, MaybeUninit<T>, and T all really just hold a T, and you can cast between them to your heart's delight, that goes out the window once you have, for example, an Option<MaybeUninit<T>>. Though Option<T> may be able to take advantage of the niche optimization (using some invalid value of T to represent None for the Option), MaybeUninit<T> can hold any bit pattern, so that optimization does not apply, and an extra byte must be kept for the Option discriminator.

It's not just optimizations that can cause layouts to diverge once wrapper types come into play. As an example, take the code in Listing 9-8; here, the layout of Wrapper<PhantomData<u8>> and Wrapper<PhantomData<i8>> is completely different even though the provided types are both empty!

```
struct Wrapper<T: SneakyTrait> {
    item: T::Sneaky,
    iter: PhantomData<T>,
}
trait SneakyTrait {
    type Sneaky;
}
impl SneakyTrait for PhantomData<u8> {
    type Sneaky = ();
}
impl SneakyTrait for PhantomData<i8> {
    type Sneaky = [u8; 1024];
}
```

Listing 9-8: Wrapper types make casting hard to get right.

All of this isn't to say that you can never cast types in Rust. Things get a lot easier, for example, when you control all of the types involved and their trait implementations, or if types are #[repr(C)]. You just need to be aware that Rust gives very few guarantees about in-memory representations, and write your code accordingly!

The Drop Check

The Rust borrow checker is, in essence, a sophisticated tool for ensuring the soundness of code at compile time, which is in turn what gives Rust a way to express code being "safe." How exactly the borrow checker does its job is beyond the scope of this book, but one check, the *drop check*, is worth going through in some detail since it has some direct implications for unsafe code. To understand drop checking, let's put ourselves in the Rust compiler's shoes for a second and look at two code snippets. First, take a look at the little three-liner in Listing 9-9 that takes a mutable reference to a variable and then mutates that same variable right after.

```
let mut x = true;
let foo = Foo(&mut x);
x = false;
```

Listing 9-9: The implementation of Foo dictates whether this code should compile

Without knowing the definition of Foo, can you say whether this code should compile or not? When we set x = false, there is still a foo hanging around that will be dropped at the end of the scope. We know that foo contains a mutable borrow of x, which would indicate that the mutable borrow that's necessary to modify x is illegal. But what's the harm in allowing it? It turns out that allowing the mutation of x is problematic only if Foo implements Drop—if Foo doesn't implement Drop, then we know that Foo won't touch the reference to x after its last use. Since that last use is before we need the exclusive reference for the assignment, we can allow the code! On the other hand, if Foo does implement Drop, we can't allow this code, since the Drop implementation may use the reference to x.

Now that you're warmed up, take a look at Listing 9-10. In this not-so-straightforward code snippet, the mutable reference is buried even deeper.

```
fn barify<'a>(_: &'a mut i32) -> Bar<Foo<'a>> { .. }
let mut x = true;
let foo = barify(&mut x);
x = false;
```

Listing 9-10: The implementations of both Foo and Bar dictate whether this code should compile

Again, without knowing the definitions of Foo and Bar, can you say whether this code should compile or not? Let's consider what happens if Foo implements Drop but Bar does not, since that's the most interesting case. Usually, when a Bar goes out of scope, or otherwise gets dropped, it'll still have to drop Foo, which in turn means that the code should be rejected for the same reason as before: Foo::drop might access the reference to x. However, Bar may not contain a Foo directly at all, but instead just a PhantomData<Foo<'a>> or a &'static Foo<'a>, in which case the code is actually okay—even though the Bar is dropped, Foo::drop is never invoked, and the reference to x is never accessed. This is the kind of code we want the compiler to accept because a human will be able to identify that it's okay, even if the compiler finds it difficult to detect that this is the case.

The logic we've just walked through is the drop check. Normally it doesn't affect unsafe code too much as its default behavior matches user expectations, with one major exception: dangling generic parameters.

Imagine that you're implementing your own Box<T> type, and someone places a &mut x into it as we did in Listing 9-9. Your Box type needs to implement Drop to free memory, but it doesn't access T beyond dropping it. Since dropping a &mut does nothing, it should be entirely fine for code to access &mut x again after the last time the Box is accessed but before it's dropped! To support types like this, Rust has an unstable feature called dropck_eyepatch (because it makes the drop check partially blind). The feature is likely to remain unstable forever and is intended to serve only as a temporary escape hatch until a proper mechanism is devised. The dropck_eyepatch feature adds a #[may_dangle] attribute, which you can add as a prefix for generic lifetimes and types in a type's Drop implementation to tell the drop check machinery that you won't use the annotated lifetime or type beyond dropping it. You use it by writing:

```
unsafe impl<#[may_dangle] T> Drop for ..
```

This escape hatch allows a type to declare that a given generic parameter isn't used in Drop, which enables use cases like Box<&mut T>. However, it also introduces a new problem if your Box<T> holds a raw heap pointer, *mut T, and allows T to dangle using #[may_dangle]. Specifically, the *mut T makes Rust's drop check think that your Box<T> doesn't own a T, and thus that it doesn't call T::drop either. Combined with the may_dangle assertion that we don't access T when the Box<T> is dropped, the drop check now concludes that it's fine to have a Box<T> where the T doesn't live until the Box is dropped (like our shortened &mut x in Listing 9-10). But that's not true, since we *do* call T::drop, which may itself access, say, a reference to said x.

Luckily, the fix is simple: we add a PhantomData<T> to tell the drop check that even though the Box<T> doesn't hold any T, and won't access T on drop, it does still own a T and will drop one when the Box is dropped. Listing 9-11 shows what our hypothetical Box type would look like.

```
struct Box<T> {
    t: NonNull<T>, // NonNull not *mut for covariance (Chapter 1)
    _owned: PhantomData<T>, // For drop check to realize we drop a T
}
unsafe impl<#[may_dangle] T> for Box<T> { /* ... */ }
```

Listing 9-11: A definition for Box that is maximally flexible in terms of the drop check

This interaction is subtle and easy to miss, but it arises only when you use the unstable #[may_dangle] attribute. Hopefully this subsection will serve as a warning so that when you see unsafe impl Drop in the wild in the future, you'll know to look for a PhantomData<T> as well!

NOTE *Another consideration for unsafe code concerning Drop is to make sure that you have a Type<T> that lets T continue to live after self is dropped. For example, if you're implementing delayed garbage collection, you need to also add T: 'static. Otherwise, if T = WriteOnDrop<&mut U>, the later access or drop of T could trigger undefined behavior!*

Coping with Fear

With this chapter mostly behind you, you may now be more afraid of unsafe code than you were before you started. While that is understandable, it's important to stress that it's not only *possible* to write safe unsafe code, but most of the time it's not even that difficult. The key is to make sure that you handle unsafe code with care; that's half the struggle. And be really sure that there isn't a safe implementation you can use instead before resorting to unsafe.

In the remainder of this chapter, we'll look at some techniques and tools that can help you be more confident in the correctness of your unsafe code when there's no way around it.

Manage Unsafe Boundaries

It's tempting to reason about unsafety *locally*; that is, to consider whether the code in the unsafe block you just wrote is safe without thinking too much about its interaction with the rest of the codebase. Unfortunately, that kind of local reasoning often comes back to bite you. A good example of this is the Unpin trait—you may write some code for your type that uses Pin::new_unchecked to produce a pinned reference to a field of the type, and that code may be entirely safe when you write it. But then at some later point in time, you (or someone else) might add a safe implementation of Unpin for said type, and suddenly the unsafe code is no longer safe, even though it's nowhere near the new impl!

Safety is a property that can be checked only at the privacy boundary of all code that relates to the unsafe block. *Privacy boundary* here isn't so much a formal term as an attempt at describing "any part of your code that can fiddle with the unsafe bits." For example, if you declare a public type Foo in a module bar that is marked pub or pub(crate), then any other code in the same crate can implement methods on and traits for Foo. So, if the safety of your unsafe code depends on Foo not implementing particular traits or methods with particular signatures, you need to remember to recheck the safety of that unsafe block any time you add an impl for Foo. If, on the other hand, Foo is not visible to the entire crate, then a much smaller set of scopes is able to add problematic implementations, and thus, the risk of accidentally adding an implementation that breaks the safety invariants goes down accordingly. If Foo is private, then only the current module and any submodules can add such implementations.

The same rule applies to access to fields: if the safety of an unsafe block depends on certain invariants over a type's fields, then any code that can touch those fields (including safe code) falls within the privacy boundary of the unsafe block. Here, too, minimizing the privacy boundary is the best approach—code that cannot get to the fields cannot mess up your invariants!

Because unsafe code often requires this wide-reaching reasoning, it's best practice to encapsulate the unsafety in your code as best you can. Provide the unsafety in the form of a single module, and strive to give that module an interface that is entirely safe. That way you only need to audit the internals of that module for your invariants. Or better yet, stick the unsafe bits in their own crate so that you can't leave any holes open by accident!

It's not always possible to fully encapsulate complex unsafe interactions to a single, safe interface, however. When that's the case, try to narrow down the parts of the public interface that have to be unsafe so that you have only a very small number of them, give them names that clearly communicate that care is needed, and then document them rigorously.

It is sometimes tempting to remove the `unsafe` marker on internal APIs so that you don't have to stick `unsafe {}` throughout your code. After all, inside your code you know never to invoke `frobnify` if you've previously called `bazzify`, right? Removing the `unsafe` annotation can lead to cleaner code but is usually a bad decision in the long run. A year from now, when your codebase has grown, you've paged out some of the safety invariants, and you "just want to hack together this one feature real quick," chances are that you'll inadvertently violate one of those invariants. And since you don't have to type `unsafe`, you won't even think to check. Plus, even if you never make mistakes, what about other contributors to your code? Ultimately, cleaner code is not a good enough argument to remove the intentionally noisy `unsafe` marker.

Read and Write Documentation

It goes without saying that if you write an unsafe function, you must document the conditions under which that function is safe to call. Here, both clarity and completeness are important. Don't leave any invariants out, even if you've already written them somewhere else. If you have a type or module that requires certain global invariants—invariants that must always hold for all uses of the type—then remind the reader that they must also uphold the global invariants in every unsafe function's documentation too. Developers often read documentation in an ad hoc, on-demand manner, so you can assume they have probably not read your carefully written module-level documentation and need to be given a nudge to do so.

What may be less obvious is that you should also document all unsafe implementations and blocks—think of this as providing proof that you do indeed uphold the contract the operation in question requires. For example, `slice::get_unchecked` requires that the provided index is within the bounds of the slice; when you call that method, put a comment just above it explaining how you know that the index is in fact guaranteed to be in bounds. In some cases, the invariants that the unsafe block requires are extensive, and your comments may get long. That's a good thing. I have caught mistakes many times by trying to write the safety comment for an unsafe block and realizing halfway through that I actually don't uphold a key invariant. You'll also thank yourself a year down the road when you have to modify this code and ensure it's still safe. And so will the contributor to your project who just stumbled across this unsafe call and wants to understand what's going on.

Before you get too deep into writing unsafe code, I also highly recommend that you go read the Rustonomicon (*https://doc.rust-lang.org/nomicon/*) cover to cover. There are so many details that are easy to miss, and will come back to bite you if you're not aware of them. We've covered many of them in this chapter, but it never hurts to be more aware. You should also make liberal use of the Rust reference whenever you're in doubt. It's added to regularly, and chances are that if you're even slightly unsure about whether some assumption you have is right, the reference will call it out. If it doesn't, consider opening an issue so that it'll be added!

Check Your Work

Okay, so you've written some unsafe code, you've double- and triple-checked all the invariants, and you think it's ready to go. Before you put it into production, there are some automated tools that you should run your test suite through (you have a test suite, right?).

The first of these is Miri, the mid-level intermediate representation interpreter. Miri doesn't compile your code into machine code but instead interprets the Rust code directly. This provides Miri with far more visibility into what your program is doing, which in turn allows it to check that your program doesn't do anything obviously bad, like read from uninitialized memory. Miri can catch a lot of very subtle and Rust-specific bugs and is a lifesaver for anyone writing unsafe code.

Unfortunately, because Miri has to interpret the code to execute it, code run under Miri often runs orders of magnitude slower than its compiled counterpart. For that reason, Miri should really be used only to execute your test suite. It can also check only the code that actually runs, and thus won't catch issues in code paths that your test suite doesn't reach. You should think of Miri as an extension of your test suite, not a replacement for it.

There are also tools known as *sanitizers*, which instrument machine code to detect erroneous behavior at runtime. The overhead and fidelity of these tools vary greatly, but one widely loved tool is Google's AddressSanitizer. It detects a large number of memory errors, such as use-after-free, buffer overflows, and memory leaks, all of which are common symptoms of incorrect unsafe code. Unlike Miri, these tools operate on machine code and thus tend to be fairly fast—usually within the same order of magnitude. But like Miri, they are constrained to analyzing the code that actually runs, so here too a solid test suite is vital.

The key to using these tools effectively is to automate them through your continuous integration pipeline so they're run for every change, and to ensure that you add regression tests over time as you discover errors. The tools get better at catching problems as the quality of your test suite improves, so by incorporating new tests as you fix known bugs, you're earning double points back, so to speak!

Finally, don't forget to sprinkle assertions generously through unsafe code. A panic is always better than triggering undefined behavior! Check all of your assumptions with assertions if you can—even things like the size of a usize if you rely on that for safety. If you're concerned about runtime cost, make use of the debug_assert* macros and the if cfg!(debug_assertions) || cfg!(test) construct to execute them only in debug and test contexts.

A HOUSE OF CARDS?

Unsafe code can violate all of Rust's safety guarantees, and this is often touted as a reason why Rust's whole safety argument is a charade. The concern is that it takes only one bit of incorrect unsafe code for the whole house to come crashing down and all safety to be lost. Proponents of this argument then sometimes argue that at the very least only unsafe code should be able to call unsafe code, so that the unsafety is visible all the way to the highest level of the application.

The argument is understandable—it is true that the safety of Rust code relies on the safety of all the transitive unsafe code it ends up invoking. And indeed, if some of that unsafe code is incorrect, it may have implications for the safety of the program overall. However, what this argument misses is that *all* successful safe languages provide a facility for language extensions that are not expressible in the (safe) surface language, usually in the form of code written in C or assembly. Just as Rust relies on the correctness of its unsafe code, the safety of those languages relies on the correctness of those extensions.

Rust is different in that it doesn't have a separate extension language, but instead allows extensions to be written in what amounts to a dialect of Rust (unsafe Rust). This allows much closer integration between the safe and unsafe code, which in turn reduces the likelihood of errors due to impedance mismatches at the interface between the two, or due to developers being familiar with one but not the other. The closer integration also makes it easier to write tools that analyze the correctness of the unsafe code's interaction with the safe code, as exemplified by tools like Miri. And since unsafe Rust continues to be subject to the borrow checker for any operation that isn't explicitly unsafe, there remain many safety checks in place that aren't present when developers must drop down to a language like C.

Summary

In this chapter, we've walked through the powers that come with the unsafe keyword and the responsibilities we accept by leveraging those powers. We also talked about the consequences of writing unsafe unsafe code, and how you really should be thinking about unsafe as a way to swear to the compiler that you've manually checked that the indicated code is still safe. In the next chapter, we'll jump into concurrency in Rust and see how you can get all those cores on your shiny new computer to pull in the same direction!

10

CONCURRENCY (AND PARALLELISM)

With this chapter I hope to provide you with all the information and tools you'll need to take effective advantage of concurrency in your Rust programs, to implement support for concurrent use in your libraries, and to use Rust's concurrency primitives correctly. I won't directly teach you how to implement a concurrent data structure or write a high-performance concurrent application. Instead, my goal is to give you sufficient understanding of the underlying mechanisms that you're equipped to wield them yourself for whatever you may need them for.

Concurrency comes in three flavors: single-thread concurrency (like with async/await, as we discussed in Chapter 8), single-core multithreaded concurrency, and multicore concurrency, which yields true parallelism.

Each flavor allows the execution of concurrent tasks in your program to be interleaved in different ways. There are even more subflavors if you take the details of operating system scheduling and preemption into account, but we won't get too deep into that.

At the type level, Rust represents only one aspect of concurrency: multithreading. Either a type is safe for use by more than one thread, or it is not. Even if your program has multiple threads (and so is concurrent) but only one core (and so is not parallel), Rust must assume that if there are multiple threads, there may be parallelism. Most of the types and techniques we'll be talking about apply equally whether two threads actually execute in parallel or not, so to keep the language simple, I'll be using the word *concurrency* in the informal sense of "things running more or less at the same time" throughout this chapter. When the distinction is important, I'll call that out.

What's particularly neat about Rust's approach to type-based safe multithreading is that it is not a feature of the compiler, but rather a library feature that developers can extend to develop sophisticated concurrency contracts. Since thread safety is expressed in the type system through Send and Sync implementations and bounds, which propagate all the way out to application code, the thread safety of the entire program is checked through type checking alone.

The Rust Programming Language already covers most of the basics when it comes to concurrency, including the Send and Sync traits, Arc and Mutex, and channels. I therefore won't reiterate much of that here, except where it's worth repeating something specifically in the context of some other topic. Instead, we'll look at what makes concurrency difficult and some common concurrency patterns intended to deal with those difficulties. We'll also explore how concurrency and asynchrony interact (and how they don't) before diving into how to use atomic operations to implement lower-level concurrent operations. Finally, I'll close out the chapter with some advice for how to retain your sanity when working with concurrent code.

The Trouble with Concurrency

Before we dive into good patterns for concurrent programming and the details of Rust's concurrency mechanisms, it's worth taking some time to understand why concurrency is challenging in the first place. That is, why do we need special patterns and mechanisms for concurrent code?

Correctness

The primary difficulty in concurrency is coordinating access—in particular, write access—to a resource that is shared among multiple threads. If lots of threads want to share a resource solely for the purposes of reading it, then that's usually easy: you stick it in an Arc or place it in something you can get a &'static to, and you're all done. But once any thread wants to write, all sorts of problems arise, usually in the form of *data races*. Briefly, a data race occurs when one thread updates shared state while a second thread is also accessing that state, either to read it or to update it. Without additional safeguards in place, the second thread may read partially overwritten state, clobber parts of what the first thread wrote, or fail to see the first thread's write at all! In general, all data races are considered undefined behavior.

Data races are a part of a broader class of problems that primarily, though not exclusively, occur in a concurrent setting: *race conditions*. A race condition occurs whenever multiple outcomes are possible from a sequence of instructions, depending on the relative timing of other events in the system. These events can be threads executing a particular piece of code, a timer going off, a network packet coming in, or any other time-variable occurrence. Race conditions, unlike data races, are not inherently bad, and are not considered undefined behavior. However, they are a breeding ground for bugs when particularly peculiar races occur, as you'll see throughout this chapter.

Performance

Often, developers introduce concurrency into their programs in the hope of increasing performance. Or, to be more precise, they hope that concurrency will enable them to perform more operations per second in aggregate by taking advantage of more hardware resources. This can be done on a single core by having one thread run while another is waiting, or across multiple cores by having threads do work simultaneously, one on each core, that would otherwise happen serially on one core. Most developers are referring to the latter kind of performance gain when they talk about concurrency, which is often framed in terms of scalability. Scalability in this context means "the performance of this program scales with the number of cores," implying that if you give your program more cores, its performance improves.

While achieving such a speedup is possible, it's harder than it seems. The ultimate goal in scalability is linear scalability, where doubling the number of cores doubles the amount of work your program completes per unit of time. Linear scalability is also often called perfect scalability. However, in reality, few concurrent programs achieve such speedups. Sublinear scaling is more common, where the throughput increases nearly linearly as you go from one core to two, but adding more cores yields diminishing returns. Some programs even experience negative scaling, where giving the program access to more cores *reduces* throughput, usually because the many threads are all contending for some shared resource.

It might help to think of a group of people trying to pop all the bubbles on a piece of bubble wrap—adding more people helps initially, but at some point you get diminishing returns as the crowding makes any one person's job harder. If the humans involved are particularly ineffective, your group may end up standing around discussing who should pop next and pop no bubbles at all! This kind of interference among tasks that are supposed to execute in parallel is called *contention* and is the archnemesis of scaling well. Contention can arise in a number of ways, but the primary offenders are mutual exclusion, shared resource exhaustion, and false sharing.

Mutual Exclusion

When only a single concurrent task is allowed to execute a particular piece of code at any one time, we say that execution of that segment of code is mutually exclusive—if one thread executes it, no other thread can do so at the same time. The archetypal example of this is a mutual exclusion lock, or *mutex*, which explicitly enforces that only one thread gets to enter a particular critical section of your program code at any one time. Mutual exclusion can also happen implicitly, however. For example, if you spin up a thread to manage a shared resource and send jobs to it over an mpsc channel, that thread effectively implements mutual exclusion, since only one such job gets to execute at a time.

Mutual exclusion can also occur when invoking operating system or library calls that internally enforce single-threaded access to a critical section. For example, for many years, the standard memory allocator required mutual exclusion for some allocations, which made memory allocation an operation that incurred significant contention in otherwise highly parallel programs. Similarly, many operating system operations that may seem like they should be independent, such as creating two files with different names in the same directory, may end up having to happen sequentially inside the kernel.

NOTE *Scalable concurrent allocations is the raison d'être for the* jemalloc *memory allocator!*

Mutual exclusion is the most obvious barrier to parallel speedup since, by definition, it forces serial execution of some portion of your program. Even if you make the remainder of your program scale with the number of cores perfectly, the total speedup you can achieve is limited by the length of the mutually exclusive, serial section. Be mindful of your mutually exclusive sections, and seek to restrict them to only where strictly necessary.

NOTE *For the theoretically minded, the limits on the achievable speedup as a result of mutually exclusive sections of code can be computed using Amdahl's law.*

Shared Resource Exhaustion

Unfortunately, even if you achieve perfect concurrency within your tasks, the environment those tasks need to interact with may itself not be perfectly scalable. The kernel can handle only so many sends on a given TCP socket per second, the memory bus can do only so many reads at once, and your GPU has a limited capacity for concurrency. There's no cure for this. The environment is usually where perfect scalability falls apart in practice, and fixes for such cases tend to require substantial re-engineering (or even new hardware!), so we won't talk much more about this topic in this chapter. Just remember that scalability is rarely something you can "achieve," and more something you just strive for.

False Sharing

False sharing occurs when two operations that shouldn't contend with one another contend anyway, preventing efficient simultaneous execution. This usually happens because the two operations happen to intersect on some shared resource even though they use unrelated parts of that resource.

The simplest example of this is lock oversharing, where a lock guards some composite state, and two operations that are otherwise independent both need to take the lock to update their particular parts of the state. This in turn means the operations must execute serially instead of in parallel. In some cases it's possible to split the single lock into two, one for each of the disjoint parts, which enables the operations to proceed in parallel. However, it's not always straightforward to split a lock like this—the state may share a single lock because some third operation needs to lock over all the parts of the state. Usually you can still split the lock, but you have to be careful about the order in which different threads take the split locks to avoid deadlocks that can occur when two operations attempt to take them in different orders (look up the "dining philosophers problem," if you're curious). Alternatively, for some problems, you may be able to avoid the critical section entirely by using a lock-free version of the underlying algorithm, though those are also tricky to get right. Ultimately, false sharing is a hard problem to solve, and there isn't a single catchall solution—but identifying the problem is a good start.

A more subtle example of false sharing occurs on the CPU level, as we discussed briefly in Chapter 2. The CPU internally operates on memory in terms of cache lines—longer sequences of consecutive bytes in memory—rather than individual bytes, to amortize the cost of memory accesses. For example, on most Intel processors, the cache line size is 64 bytes. This means that every memory operation really ends up reading or writing some multiple of 64 bytes. The false sharing comes into play when two cores want to update the value of two different bytes that happen to fall on the same cache line; those updates must execute sequentially even though the updates are logically disjoint.

This might seem too low-level to matter, but in practice this kind of false sharing can decimate the parallel speedup of an application. Imagine that you allocate an array of integer values to indicate how many operations each thread has completed, but the integers all fall within the same cache line—now, all your otherwise parallel threads will contend on that one cache line for every operation they perform. If the operations are relatively quick, *most* of your execution time may end up being spent contending on those counters!

The trick to avoiding false cache line sharing is to pad your values so that they are the size of a cache line. That way, two adjacent values always fall on different cache lines. But of course, this also inflates the size of your data structures, so use this approach only when benchmarks indicate a problem.

Concurrency Models

Rust has three patterns for adding concurrency to your programs that you'll come across fairly often: shared memory concurrency, worker pools, and actors. Going through every way you could add concurrency in detail would take a book of its own, so here I'll focus on just these three patterns.

Shared Memory

Shared memory concurrency is, conceptually, very straightforward: the threads cooperate by operating on regions of memory shared between them. This might take the form of state guarded by a mutex or stored in a hash map with support for concurrent access from many threads. The many threads may be doing the same task on disjoint pieces of data, such as if many threads perform some function over disjoint subranges of a Vec, or they may be performing different tasks that require some shared state, such as in a database where one thread handles user queries to a table while another optimizes the data structures used to store that table in the background.

When you use shared memory concurrency, your choice of data structures is significant, especially if the threads involved need to cooperate very closely. A regular mutex might prevent scaling beyond a very small number of cores, a reader/writer lock might allow many more concurrent reads at the cost of slower writes, and a sharded reader/writer lock might allow perfectly scalable reads at the cost of making writes highly disruptive. Similarly, some concurrent hash maps aim for good all-round performance while others specifically target, say, concurrent reads where writes are rare. In general, in shared memory concurrency, you want to use data structures that are specifically designed for something as close to your target use case as possible, so that you can take advantage of optimizations that trade off performance aspects your application does not care about for those it does.

Shared memory concurrency is a good fit for use cases where threads need to jointly update some shared state in a way that does not commute. That is, if one thread has to update the state s with some function f, and another has to update the state with some function g, and f(g(s)) != g(f(s)), then shared memory concurrency is likely necessary. If that is not the case, the other two patterns are likely better fits, as they tend to lead to simpler and more performant designs.

NOTE *Some problems have known algorithms that can provide concurrent shared memory operations without the use of locks. As the number of cores grows, these lock-free algorithms may scale better than lock-based algorithms, though they also often have slower per-core performance due to their complexity. As always with performance matters, benchmark first, then look for alternative solutions.*

Worker Pools

In the worker pool model, many identical threads receive jobs from a shared job queue, which they then execute entirely independently. Web servers, for example, often have a worker pool handling incoming connections, and multithreaded runtimes for asynchronous code tend to use a worker pool to collectively execute all of an application's futures (or, more accurately, its top-level tasks).

The lines between shared memory concurrency and worker pools are often blurry, as worker pools tend to use shared memory concurrency to coordinate how they take jobs from the queue and how they return incomplete jobs back to the queue. For example, say you're using the data parallelism library rayon to perform some function over every element of a vector in parallel. Behind the scenes rayon spins up a worker pool, splits the vector into subranges, and then hands out subranges to the threads in the pool. When a thread in the pool finishes a range, rayon arranges for it to start working on the next unprocessed subrange. The vector is shared among all the worker threads, and the threads coordinate through a shared memory queue–like data structure that supports work stealing.

Work stealing is a key feature of most worker pools. The basic premise is that if one thread finishes its work early, and there's no more unassigned work available, that thread can steal jobs that have already been assigned to a different worker thread but haven't been started yet. Not all jobs take the same amount of time to complete, so even if every worker is given the same *number* of jobs, some workers may end up finishing their jobs more quickly than others. Rather than sit around and wait for the threads that drew longer-running jobs to complete, those threads that finish early should help the stragglers so the overall operation is completed sooner.

It's quite a task to implement a data structure that supports this kind of work stealing without incurring significant overhead from threads constantly trying to steal work from one another, but this feature is vital to a high-performance worker pool. If you find yourself in need of a worker pool, your best bet is usually to use one that has already seen a lot of work go into it, or at least reuse data structures from an existing one, rather than to write one yourself from scratch.

Worker pools are a good fit when the work that each thread performs is the same, but the data it performs it *on* varies. In a rayon parallel map operation, every thread performs the same map computation; they just perform it on different subsets of the underlying data. In a multithreaded asynchronous runtime, each thread simply calls Future::poll; they just call it on different futures. If you start having to distinguish between the threads in your thread pool, a different design is probably more appropriate.

CONNECTION POOLS

A connection pool is a shared memory construct that keeps a set of established connections and hands them out to threads that need a connection. It's a common design pattern in libraries that manage connections to external services. If a thread needs a connection but one isn't available, either a new connection is established or the thread is forced to block. When a thread is done with a connection, it returns that connection to the pool, and thus makes it available to other threads that may be waiting.

Usually, the hardest task for a connection pool is managing connection life cycles. A connection can be returned to the pool in whatever state it was put in by the last thread that used it. The connection pool therefore has to make sure any state associated with the connection, whether on the client or on the server, has been reset so that when the connection is subsequently used by another thread, that thread can act as though it was given a fresh, dedicated connection.

Actors

The actor concurrency model is, in many ways, the opposite of the worker pool model. Whereas the worker pool has many identical threads that share a job queue, the actor model has many separate job queues, one for each job "topic." Each job queue feeds into a particular actor, which handles all jobs that pertain to a subset of the application's state. That state might be a database connection, a file, a metrics collection data structure, or any other structure that you can imagine many threads may need to be able to access. Whatever it is, a single actor owns that state, and if some task wants to interact with that state, it needs to send a message to the owning actor summarizing the operation it wishes to perform. When the owning actor receives that message, it performs the indicated action and responds to the inquiring task with the result of the operation, if relevant.

Since the actor has exclusive access to its inner resource, no locks or other synchronization mechanisms are required beyond what's needed for the messaging.

A key point in the actor pattern is that actors all talk to one another. If, say, an actor that is responsible for logging needs to write to a file and a database table, it might send off messages to the actors responsible for each of those, asking them to perform the respective actions, and then proceed to the next log event. In this way, the actor model more closely resembles a web than spokes on a wheel—a user request to a web server might start as a single request to the actor responsible for that connection but might transitively spawn tens, hundreds, or even thousands of messages to actors deeper in the system before the user's request is satisfied.

Nothing in the actor model requires that each actor is its own thread. To the contrary, most actor systems suggest that there should be a large number of actors, and so each actor should map to a task rather than a thread. After all, actors require exclusive access to their wrapped resources only when they execute, and do not care whether they are on a thread of their own or not. In fact, very frequently, the actor model is used in conjunction with the worker pool model—for example, an application that uses the multi-threaded asynchronous runtime Tokio can spawn an asynchronous task for each actor, and Tokio will then make the execution of each actor a job in its worker pool. Thus, the execution of a given actor may move from thread to thread in the worker pool as the actor yields and resumes, but every time the actor executes it maintains exclusive access to its wrapped resource.

The actor concurrency model is well suited for when you have many resources that can operate relatively independently, and where there is little or no opportunity for concurrency within each resource. For example, an operating system might have an actor responsible for each hardware device, and a web server might have an actor for each backend database connection. The actor model does not work so well if you need only a few actors, if work is skewed significantly among the actors, or if some actors grow large—in all of those cases, your application may end up being bottle-necked on the execution speed of a single actor in the system. And since actors each expect to have exclusive access to their little slice of the world, you can't easily parallelize the execution of that one bottleneck actor.

Asynchrony and Parallelism

As we discussed in Chapter 8, asynchrony in Rust enables concurrency without parallelism—we can use constructs like selects and joins to have a single thread poll multiple futures and continue when one, some, or all of them complete. Because there is no parallelism involved, concurrency with futures does not fundamentally require those futures to be Send. Even spawning a future to run as an additional top-level task does not fundamentally require Send, since a single executor thread can manage the polling of many futures at once.

However, in *most* cases, applications want both concurrency and parallelism. For example, if a web application constructs a future for each incoming connection and so has many active connections at once, it probably wants the asynchronous executor to be able to take advantage of more than one core on the host computer. That won't happen naturally: your code has to explicitly tell the executor which futures can run in parallel and which cannot.

In particular, two pieces of information must be given to the executor to let it know that it can spread the work in the futures across a worker pool of threads. The first is that the futures in question are Send—if they aren't, the executor is not allowed to send the futures to other threads for processing, and no parallelism is possible; only the thread that constructed such futures can poll them.

The second piece of information is how to split the futures into tasks that can operate independently. This ties back to the discussion of tasks versus futures from Chapter 8: if one giant Future contains a number of Future instances that themselves correspond to tasks that can run in parallel, the executor must still call poll on the top-level Future, and it must do so from a single thread, since poll requires &mut self. Thus, to achieve parallelism with futures, you have to explicitly spawn the futures you want to be able to run in parallel. Also, because of the first requirement, the executor function you use to do so will require that the passed-in Future is Send.

ASYNCHRONOUS SYNCHRONIZATION PRIMITIVES

Most of the synchronization primitives that exist for blocking code (think std::sync) also have asynchronous counterparts. There are asynchronous variants of channels, mutexes, reader/writer locks, barriers, and all sorts of other similar constructs. We need these because, as discussed in Chapter 8, blocking inside a future will hold up other work the executor may need to do, and so is inadvisable.

However, the asynchronous versions of these primitives are often slower than their synchronous counterparts because of the additional machinery needed to perform the necessary wake-ups. For that reason, you may want to use synchronous synchronization primitives even in asynchronous contexts whenever the use does not risk blocking the executor. For example, while it's generally true that acquiring a Mutex might block for a long time, that might not be true for a particular Mutex that, perhaps, is acquired only rarely, and only ever for short periods of time. In that case, blocking for the short time until the Mutex becomes available again might not actually cause any problems. You will want to make sure that you never yield or perform other long-running operations while holding the MutexGuard, but barring that you shouldn't run into problems.

As always with such optimizations, though, make sure you measure first, and choose only the synchronous primitive if it nets you significant performance improvements. If it does not, the additional footguns introduced by using a synchronous primitive in an asynchronous context are probably not worth it.

Lower-Level Concurrency

The standard library provides the std::sync::atomic module, which provides access to the underlying CPU primitives, higher-level constructs like channels and mutexes are built with. These primitives come in the form of atomic types with names starting with Atomic—AtomicUsize, AtomicI32, AtomicBool, AtomicPtr, and so on—the Ordering type, and two functions called fence and compiler_fence. We'll look at each of these over the next few sections.

These types are the blocks used to build any code that has to communicate between threads. Mutexes, channels, barriers, concurrent hash tables, lock-free stacks, and all other synchronization constructs ultimately rely on these few primitives to do their jobs. They also come in handy on their own for lightweight cooperation between threads where heavyweight synchronization like a mutex is excessive—for example, to increment a shared counter or set a shared Boolean to true.

The atomic types are special in that they have defined semantics for what happens when multiple threads try to access them concurrently. These types all support (mostly) the same API: load, store, fetch_*, and compare_exchange. In the rest of this section, we'll look at what those do, how to use them correctly, and what they're useful for. But first, we have to talk about low-level memory operations and memory ordering.

Memory Operations

Informally, we often refer to accessing variables as "reading from" or "writing to" memory. In reality, a lot of machinery between code uses a variable and the actual CPU instructions that access your memory hardware. It's important to understand that machinery, at least at a high level, in order to understand how concurrent memory accesses behave.

The compiler decides what instructions to emit when your program reads the value of a variable or assigns a new value to it. It is permitted to perform all sorts of transformations and optimizations on your code and may end up reordering your program statements, eliminating operations it deems redundant, or using CPU registers rather than actual memory to store intermediate computations. The compiler is subject to a number of restrictions on these transformations, but ultimately only a subset of your variable accesses actually end up as memory access instructions.

At the CPU level, memory instructions come in two main shapes: loads and stores. A load pulls bytes from a location in memory into a CPU register, and a store stores bytes from a CPU register into a location in memory. Loads and stores operate on small chunks of memory at a time: usually 8 bytes or less on modern CPUs. If a variable access spans more bytes than can be accessed with a single load or store, the compiler automatically turns it into multiple load or store instructions, as appropriate. The CPU also has some leeway in how it executes a program's instructions to make better use of the hardware and improve program performance. For example, modern CPUs often execute instructions in parallel, or even out of order, when they don't have dependencies on each other. There are also several layers of caches between each CPU and your computer's DRAM, which means that a load of a given memory location may not necessarily see the latest store to that memory location, going by wall-clock time.

In most code, the compiler and CPU are permitted to transform the code only in ways that don't affect the semantics of the resulting program, so these transformations are invisible to the programmer. However, in the context of parallel execution, these transformations can have a significant impact on application behavior. Therefore, CPUs typically provide multiple different variations of the load and store instructions, each with different guarantees about how the CPU may reorder them and how they may be interleaved with parallel operations on other CPUs. Similarly, compilers (or rather, the language the compiler compiles) provide different annotations you can use to force particular execution constraints for some subset of their memory accesses. In Rust, those annotations come in the form of the atomic types and their methods, which we'll spend the rest of this section picking apart.

Atomic Types

Rust's atomic types are so called because they can be accessed atomically—that is, the value of an atomic-type variable is written all at once and will never be written using multiple stores, guaranteeing that a load of that variable cannot observe that only some of the bytes composing the value have changed while others have not (yet). This is easiest understood by way of contrast with non-atomic types. For example, reassigning a new value to a tuple of type (i64, i64) typically requires two CPU store instructions, one for each 8-byte value. If one thread were to perform both of those stores, another thread could (if we ignore the borrow checker for a second) read the tuple's value after the first store but before the second, and thus end up with an inconsistent view of the tuple's value. It would end up reading the new value for the first element and the old value for the second element, a value that was never actually stored by any thread.

The CPU can atomically access values only of certain sizes, so there are only a few atomic types, all of which live in the atomic module. Each atomic type is of one of the sizes the CPU supports atomic access to, with multiple variations for things like whether the value is signed and to differentiate between an atomic usize and a pointer (which is of the same size as usize). Furthermore, the atomic types have explicit methods for loading and storing the values they hold, and a handful of more complex methods we'll get back to later, so that the mapping between the code the programmer writes and the resulting CPU instructions is clearer. For example, AtomicI32::load performs a single load of a signed 32-bit value, and AtomicPtr::store performs a single store of a pointer-sized (64 bits on a 64-bit platform) value.

Memory Ordering

Most of the methods on the atomic types take an argument of type Ordering, which dictates the memory ordering restrictions the atomic operation is subject to. Across different threads, loads and stores of an atomic value may be sequenced by the compiler and CPU only in interleavings that are compatible with the requested memory ordering of each of the atomic operations on that atomic value. Over the next few sections, we'll see some examples of why control over the ordering is important and necessary to get the expected semantics out of the compiler and CPU.

Memory ordering often comes across as counterintuitive, because we humans like to read programs from top to bottom and imagine that they execute line by line—but that's not how the code actually executes when it hits the hardware. Memory accesses can be reordered, or even entirely elided, and writes on one thread may not immediately be visible to other threads, even if later writes in program order have already been observed.

Think of it like this: each memory location sees a sequence of modifications coming from different threads, and the sequences of modifications for different memory locations are independent. If two threads T1 and T2 both write to memory location M, then even if T1 executed first as measured by a user with a stopwatch, T2's write to M may still appear to have happened first for M absent any other constraints between the two threads' execution. Essentially, *the computer does not take wall-clock time into account* when it determines the value of a given memory location—all that matter are the execution constraints the programmer puts on what constitutes a valid execution. For example, if T1 writes to M and then spawns thread T2, which then writes to M, the computer must recognize T1's write as having happened first because T2's existence depends on T1.

If that's hard to follow, don't fret—memory ordering can be mind-bending, and language specifications tend to use very precise but not very intuitive wording to describe it. We can construct a mental model that's easier to grasp, if a little simplified, by instead focusing on the underlying hardware architecture. Very basically, your computer memory is structured as a treelike hierarchy of storage where the leaves are CPU registers and the roots are the storage on your physical memory chips, often called main memory. Between the two are several layers of caches, and different layers of the hierarchy can reside on different pieces of hardware. When a thread performs a store to a memory location, what really happens is that the CPU starts a write request for the value in a given CPU register that then has to make its way up the memory hierarchy toward main memory. When a thread performs a load, the request flows up the hierarchy until it hits a layer that has the value available, and returns from there. Herein lies the problem: writes aren't visible everywhere until all caches of the written memory location have been updated, but other CPUs can execute instructions against the same memory location at the same time, and weirdness ensues. Memory ordering, then, is a way to request precise semantics for what happens when multiple CPUs access a particular memory location for a particular operation.

With this in mind, let's take a look at the Ordering type, which is the primary mechanism by which we, as programmers, can dictate additional constraints on what concurrent executions are valid.

Ordering is defined as an enum with the variants shown in Listing 10-1.

```
enum Ordering {
    Relaxed,
    Release,
    Acquire,
    AcqRel,
    SeqCst
}
```

Listing 10-1: The definition of Ordering

Each of these places different restrictions on the mapping from source code to execution semantics, and we'll explore each one in turn in the remainder of this section.

Relaxed Ordering

Relaxed ordering essentially guarantees nothing about concurrent access to the value beyond the fact that the access is atomic. In particular, relaxed ordering gives no guarantees about the relative ordering of memory accesses across different threads. This is the weakest form of memory ordering. Listing 10-2 shows a simple program in which two threads access two atomic variables using Ordering::Relaxed.

```
static X: AtomicBool = AtomicBool::new(false);
static Y: AtomicBool = AtomicBool::new(false);

let t1 = spawn(|| {
  ❶ let r1 = Y.load(Ordering::Relaxed);
  ❷ X.store(r1, Ordering::Relaxed);
});
let t2 = spawn(|| {
  ❸ let r2 = X.load(Ordering::Relaxed);
  ❹ Y.store(true, Ordering::Relaxed)
});
```

Listing 10-2: Two racing threads with Ordering::Relaxed

Looking at the thread spawned as t2, you might expect that r2 can never be true, since all values are false until the same thread assigns true to Y on the line *after* reading X. However, with a relaxed memory ordering, that outcome is completely possible. The reason is that the CPU is allowed to reorder the loads and stores involved. Let's walk through exactly what happens here to make r2 = true possible.

First, the CPU notices that ❹ doesn't have to happen after ❸, since ❹ doesn't use any output or side effect of ❸. That is, ❹ has no execution dependency on ❸. So, the CPU decides to reorder them for *waves hands* reasons that'll make your program go faster. The CPU thus goes ahead and executes ❹ first, setting Y = true, even though ❸ hasn't run yet. Then, t2 is put to sleep by the operating system and thread t1 executes a few instructions, or t1 simply executes on another core. In t1, the compiler must indeed run ❶ first and then ❷, since ❷ depends on the value read in ❶. Therefore, t1 reads true from

Y (written by ❹) into r1 and then writes that back to X. Finally, t2 executes ❸, which reads X and gets true, as was written by ❷.

The relaxed memory ordering allows this execution because it imposes no additional constraints on concurrent execution. That is, under relaxed memory ordering, the compiler must ensure only that execution dependencies on any given thread are respected (just as if atomics weren't involved); it need not make any promises about the interleaving of concurrent operations. Reordering ❸ and ❹ is permitted for a single-threaded execution, so it is permitted under relaxed ordering as well.

In some cases, this kind of reordering is fine. For example, if you have a counter that just keeps track of metrics, it doesn't really matter when exactly it executes relative to other instructions, and Ordering::Relaxed is fine. In other cases, this could be disastrous: say, if your program uses r2 to figure out if security protections have already been set up, and thus ends up erroneously believing that they already have been.

You don't generally notice this reordering when writing code that doesn't make fancy use of atomics—the CPU has to promise that there is no observable difference between the code as written and what each thread actually executes, so everything seems like it runs in order just as you wrote it. This is referred to as respecting program order or evaluation order; the terms are synonyms.

Acquire/Release Ordering

At the next step up in the memory ordering hierarchy, we have Ordering::Acquire, Ordering::Release, and Ordering::AcqRel (acquire plus release). At a high level, these establish an execution dependency between a store in one thread and a load in another and then restrict how operations can be reordered with respect to that load and store. Crucially, these dependencies not only establish a relationship between a store and a load of a single value, but also put ordering constraints on *other* loads and stores in the threads involved. This is because every execution must respect the program order; if a load in thread B has a dependency on some store in thread A (the store in A must execute before the load in B), then any read or write in B after that load must also happen after that store in A.

NOTE *The Acquire memory ordering can be applied only to loads, Release only to stores, and AcqRel only to operations that both load and store (like fetch_add).*

Concretely, these memory orderings place the following restrictions on execution:

1. Loads and stores cannot be moved forward past a store with Ordering::Release.

2. Loads and stores cannot be moved back before a load with Ordering::Acquire.

3. An Ordering::Acquire load of a variable must see all stores that happened before an Ordering::Release store that stored what the load loaded.

To see how these memory orderings change things, Listing 10-3 shows Listing 10-2 again but with the memory ordering swapped out for `Acquire` and `Release`.

```
static X: AtomicBool = AtomicBool::new(false);
static Y: AtomicBool = AtomicBool::new(false);

let t1 = spawn(|| {
    let r1 = Y.load(Ordering::Acquire);
    X.store(r1, Ordering::Release);
});
let t2 = spawn(|| {
  ❶ let r2 = X.load(Ordering::Acquire);
  ❷ Y.store(true, Ordering::Release)
});
```

Listing 10-3: Listing 10-2 with Acquire/Release memory ordering

These additional restrictions mean that it is no longer possible for t2 to see r2 = true. To see why, consider the primary cause of the weird outcome in Listing 10-2: the reordering of ❶ and ❷. The very first restriction, on stores with `Ordering::Release`, dictates that we cannot move ❶ below ❷, so we're all good!

But these rules are useful beyond this simple example. For example, imagine that you implement a mutual exclusion lock. You want to make sure that any loads and stores a thread runs while it holds the lock are executed only while it's actually holding the lock, and visible to any thread that takes the lock later. This is exactly what `Release` and `Acquire` enable you to do. By performing a `Release` store to release the lock and an `Acquire` load to acquire the lock, you can guarantee that the loads and stores in the critical section are never moved to before the lock was actually acquired or to after the lock was released!

NOTE   *On some CPU architectures, like x86, `Acquire/Release` ordering is guaranteed by the hardware, and there is no additional cost to using `Ordering::Release` and `Ordering::Acquire` over `Ordering::Relaxed`. On other architectures that is not the case, and your program may see speedups if you switch to `Relaxed` for atomic operations that can tolerate the weaker memory ordering guarantees.*

Sequentially Consistent Ordering

Sequentially consistent ordering (`Ordering::SeqCst`) is the strongest memory ordering we have access to. Its exact guarantees are somewhat hard to nail down, but very broadly, it requires not only that each thread sees results consistent with `Acquire/Release`, but also that all threads see the *same* ordering as one another. This is best seen by way of contrast with the behavior of `Acquire` and `Release`. Specifically, `Acquire/Release` ordering does *not* guarantee that if two threads A and B atomically load values written by two other threads X and Y, A and B will see a consistent pattern of when X wrote relative to Y. That's fairly abstract, so consider the example in Listing 10-4, which shows a case where Acquire/Release ordering can produce unexpected results. Afterwards, we'll see how sequentially consistent ordering avoids that particular unexpected outcome.

```
static X: AtomicBool = AtomicBool::new(false);
static Y: AtomicBool = AtomicBool::new(false);
static Z: AtomicI32 = AtomicI32::new(0);

let t1 = spawn(|| {
    X.store(true, Ordering::Release);
});
let t2 = spawn(|| {
    Y.store(true, Ordering::Release);
});
let t3 = spawn(|| {
    while (!X.load(Ordering::Acquire)) {}
  ❶ if (Y.load(Ordering::Acquire)) {
        Z.fetch_add(1, Ordering::Relaxed); }
});
let t4 = spawn(|| {
    while (!Y.load(Ordering::Acquire)) {}
  ❷ if (X.load(Ordering::Acquire)) {
        Z.fetch_add(1, Ordering::Relaxed); }
});
```

Listing 10-4: Weird results with Acquire/Release ordering

The two threads t1 and t2 set X and Y to true, respectively. Thread t3 waits for X to be true; once X is true, it checks if Y is true and, if so, adds 1 to Z. Thread t4 instead waits for Y to become true, and then checks if X is true and, if so, adds 1 to Z. At this point the question is: what are the possible values for Z after all the threads terminate? Before I show you the answer, try to work your way through it given the definitions of Release and Acquire ordering in the previous section.

First, let's recap the conditions under which Z is incremented. Thread t3 increments Z if it sees that Y is true after it observes that X is true, which can happen only if t2 runs before t3 evaluates the load at ❶. Conversely, thread t4 increments Z if it sees that X is true after it observes that Y is true, so only if t1 runs before t4 evaluates the load at ❷. To simplify the explanation, let's assume for now that each thread runs to completion once it runs.

Logically, then, Z can be incremented twice if the threads run in the order 1, 2, 3, 4—both X and Y are set to true, and then t3 and t4 run to find that their conditions for incrementing Z are met. Similarly, Z can trivially be incremented just once if the threads run in the order 1, 3, 2, 4. This satisfies t4's condition for incrementing Z, but not t3's. Getting Z to be 0, however, *seems* impossible: if we want to prevent t3 from incrementing Z, t2 has to run after t3. Since t3 runs only after t1, that implies that t2 runs after t1. However, t4 won't run until after t2 has run, so t1 must have run and set X to true by the time t4 runs, and so t4 will increment Z.

Our inability to get Z to be 0 stems mostly from our human inclination for linear explanations; this happened, then this happened, then this happened. Computers aren't limited in the same way and have no need to box all events into a single global order. There's nothing in the rules for Release and Acquire that says that t3 must observe the same execution order for t1 and t2 as t4 observes. As far as the computer is concerned, it's fine to let t3 observe t1 as having executed first, while having t4 observe t2 as having executed first. With that in mind, an execution in which t3 observes that Y is false after it observes that X is true (implying that t2 runs after t1), while in the same execution t4 observes that X is false after it observes that Y is true (implying that t2 runs before t1), is completely reasonable, even if that seems outrageous to us mere humans.

As we discussed earlier, Acquire/Release requires only that an Ordering::Acquire load of a variable must see all stores that happened before an Ordering::Release store that stored what the load loaded. In the ordering just discussed, the computer *did* uphold that property: t3 sees X == true, and indeed sees all stores by t1 prior to it setting X = true—there are none. It also sees Y == false, which was stored by the main thread at program startup, so there aren't any relevant stores to be concerned with. Similarly, t4 sees Y = true and also sees all stores by t2 prior to setting Y = true—again, there are none. It also sees X == false, which was stored by the main thread and has no preceding store. No rules are broken, yet it just seems wrong somehow.

Our intuitive expectation was that we could put the threads in some global order to make sense of what every thread saw and did, but that was not the case for Acquire/Release ordering in this example. To achieve something closer to that intuitive expectation, we need sequential consistency. Sequential consistency requires all the threads taking part in an atomic operation to coordinate to ensure that what each thread observes corresponds to (or at least appears to correspond to) *some* single, common execution order. This makes it easier to reason about but also makes it costly.

Atomic loads and stores marked with Ordering::SeqCst instruct the compiler to take any extra precautions (such as using special CPU instructions) needed to guarantee sequential consistency for those loads and stores. The exact formalism around this is fairly convoluted, but sequential consistency essentially ensures that if you looked at all the related SeqCst operations from across all your threads, you could put the thread executions in *some* order so that the values that were loaded and stored would all match up.

If we replaced all the memory ordering arguments in Listing 10-4 with SeqCst, Z could not possibly be 0 after all the threads have exited, just as we originally expected. Under sequential consistency, it must be possible to say either that t1 definitely ran before t2 or that t2 definitely ran before t1, so the execution where t3 and t4 see different orders is not allowed, and thus Z cannot be 0.

Compare and Exchange

In addition to load and store, all of Rust's atomic types provide a method called compare_exchange. This method is used to atomically *and conditionally* replace a value. You provide compare_exchange with the last value you observed for an atomic variable and the new value you want to replace the original value with, and it will replace the value only if it is still the same as it was when you last observed it. To see why this is important, take a look at the (broken) implementation of a mutual exclusion lock in Listing 10-5. This implementation keeps track of whether the lock is held in the static atomic variable LOCK. We use the Boolean value true to represent that the lock is held. To acquire the lock, a thread waits for LOCK to be false, then sets it to true again; it then enters its critical section and sets LOCK to false to release the lock when its work (f) is done.

```
static LOCK: AtomicBool = AtomicBool::new(false);

fn mutex(f: impl FnOnce()) {
    // Wait for the lock to become free (false).
    while LOCK.load(Ordering::Acquire)
      { /* .. TODO: avoid spinning .. */ }
    // Store the fact that we hold the lock.
    LOCK.store(true, Ordering::Release);
    // Call f while holding the lock.
    f();
    // Release the lock.
    LOCK.store(false, Ordering::Release);
}
```

Listing 10-5: An incorrect implementation of a mutual exclusion lock

This mostly works, but it has a terrible flaw—two threads might both see LOCK == false at the same time and both leave the while loop. Then they both set LOCK to true and both enter the critical section, which is exactly what the mutex function was supposed to prevent!

The issue in Listing 10-5 is that there is a gap between when we load the current value of the atomic variable and when we subsequently update it, during which another thread might get to run and read or touch its value. It is exactly this problem that compare_exchange solves—it swaps out the value behind the atomic variable *only* if its value still matches the previous read, and otherwise notifies you that the value has changed. Listing 10-6 shows the corrected implementation using compare_exchange.

```
static LOCK: AtomicBool = AtomicBool::new(false);

fn mutex(f: impl FnOnce()) {
    // Wait for the lock to become free (false).
    loop {
        let take = LOCK.compare_exchange(
            false,
            true,
            Ordering::AcqRel,
            Ordering::Relaxed
        );
        match take {
            Ok(false) => break,
            Ok(true) | Err(false) => unreachable!(),
```

```
            Err(true) => { /* .. TODO: avoid spinning .. */ }
        }
    }
    // Call f while holding the lock.
    f();
    // Release the lock.
    LOCK.store(false, Ordering::Release);
}
```

Listing 10-6: A corrected implementation of a mutual exclusion lock

This time around, we use `compare_exchange` in the loop, and it takes care of both checking that the lock is currently not held and storing true to take the lock as appropriate. This happens through the first and second arguments to `compare_exchange`, respectively: in this case, false and then true. You can read the invocation as "Store true only if the current value is false." The `compare_exchange` method returns a Result that indicates either that the value was successfully updated (Ok) or that it could not be updated (Err). In either case, it also returns the current value. This isn't too useful with an AtomicBool since we know what the value must be if the operation failed, but for something like an AtomicI32, the updated current value will let you quickly recompute what to store and then try again without having to do another load.

NOTE *Note that `compare_exchange` checks only whether the value is the same as the one that was passed in as the current value. If some other thread modifies the atomic variable's value and then resets it to the original value again, a `compare_exchange` on that variable will still succeed. This is often referred to as the A-B-A problem.*

Unlike simple loads and stores, `compare_exchange` takes *two* Ordering arguments. The first is the "success ordering," and it dictates what memory ordering should be used for the load and store that the `compare_exchange` represents in the case that the value was successfully updated. The second is the "failure ordering," and it dictates the memory ordering for the load if the loaded value does not match the expected current value. These two orderings are kept separate so that the developer can give the CPU leeway to improve execution performance by reordering loads and stores on failure when appropriate, but still get the correct ordering on success. In this case, it's okay to reorder loads and stores across failed iterations of the lock acquisition loop, but it's *not* okay to reorder loads and stores inside the critical section in such a way that they end up outside of it.

Even though its interface is simple, `compare_exchange` is a very powerful synchronization primitive—so much so that it's been theoretically proven that you can build all other distributed consensus primitives using only `compare_exchange`! For that reason, it is the workhorse of many, if not most, synchronization constructs when you really dig into the implementation details.

Be aware, though, that a `compare_exchange` requires that a single CPU has exclusive access to the underlying value, and it is therefore a form of mutual exclusion at the hardware level. This in turn means that `compare_exchange` can quickly become a scalability bottleneck: only one CPU can make progress at a time, so there's a portion of your code that will not scale with the number of cores. In fact, it's probably worse than that—the CPUs have to coordinate to ensure that only one CPU succeeds at a compare_exchange for a variable at a time (take a look at the MESI protocol if you're curious about how that works), and that coordination grows quadratically more costly the more CPUs are involved!

COMPARE_EXCHANGE_WEAK

The careful documentation reader will notice that compare_exchange has a suspiciously named cousin, compare_exchange_weak, and wonder what the difference is. The weak variant of compare_exchange is allowed to fail even if the atomic variable's value does still match the expected value that the user passed in, whereas the strong variant must succeed in this case.

This might seem odd—how could an atomic value swap fail except if the value has changed? The answer lies in system architectures that do not have a native compare_exchange operation. For example, ARM processors instead have *locked load* and *conditional store* operations, where a conditional store will fail if the value read by an associated locked load has not been written to since the load. The Rust standard library implements compare_exchange on ARM by calling this pair of instructions in a loop and returning only once the conditional store succeeds. This makes the code in Listing 10-6 needlessly inefficient—we end up with a nested loop, which requires more instructions and is harder to optimize. Since we already have a loop in this case, we could instead use compare_exchange_weak, remove the unreachable!() on Err(false), and get better machine code on ARM and the same compiled code on x86!

The Fetch Methods

Fetch methods (fetch_add, fetch_sub, fetch_and, and the like) are designed to allow more efficient execution of atomic operations that commute—that is, operations that have meaningful semantics regardless of the order they execute in. The motivation for this is that the compare_exchange method is powerful, but also costly—if two threads both want to update a single atomic variable, one will succeed, while the other will fail and have to retry. If many threads are involved, they all have to mediate sequential access to the underlying value, and there will be plenty of spinning while threads retry on failure.

For simple operations that commute, rather than fail and retry just because another thread modified the value, we can tell the CPU what operation to perform on the atomic variable. It'll then perform that operation on whatever the current value happens to be when the CPU eventually gets exclusive access. Think of an AtomicUsize that counts the number of operations a pool of threads has completed. If two threads both complete a job at the same time, it doesn't matter which one updates the counter first as long as both their increments are counted.

The fetch methods implement these kinds of commutative operations. They perform a read *and* a store operation in a single step and guarantee that the store operation was performed on the atomic variable when it held exactly the value returned by the method. As an example, `AtomicUsize::fetch_add(1, Ordering::Relaxed)` never fails—it always adds 1 to the current value of the `AtomicUsize`, no matter what it is, and returns the value of the `AtomicUsize` precisely when this thread's 1 was added.

The fetch methods tend to be more efficient than `compare_exchange` because they don't require threads to fail and retry when multiple threads contend for access to a variable. Some hardware architectures even have specialized fetch method implementations that scale much better as the number of involved CPUs grows. Nevertheless, if enough threads try to operate on the same atomic variable, those operations will begin to slow down and exhibit sublinear scaling due to the coordination required. In general, the best way to significantly improve the performance of a concurrent algorithm is to split contended variables into more atomic variables that are each less contended, rather than switching from `compare_exchange` to a fetch method.

NOTE *The* `fetch_update` *method is somewhat deceptively named—behind the scenes, it is really just a* `compare_exchange_weak` *loop, so its performance profile will more closely match that of* `compare_exchange` *than the other fetch methods.*

Sane Concurrency

Writing correct and performant concurrent code is harder than writing sequential code; you have to consider not only possible execution interleavings but also how your code interacts with the compiler, the CPU, and the memory subsystem. With such a wide array of footguns at your disposal, it's easy to want to throw your hands in the air and just give up on concurrency altogether. In this section we'll explore some techniques and tools that can help ensure that you write correct concurrent code without (as much) fear.

Start Simple

It is a fact of life that simple, straightforward, easy-to-follow code is more likely to be correct. This principle also applies to concurrent code—always start with the simplest concurrent design you can think of, then measure, and only if measurement reveals a performance problem should you optimize your algorithm.

To follow this tip in practice, start out with concurrency patterns that do not require intricate use of atomics or lots of fine-grained locks. Begin with multiple threads that run sequential code and communicate over channels, or that cooperate through locks, and then benchmark the resulting performance with the workload you care about. You're much less likely to make mistakes this way than by implementing fancy lockless algorithms or by splitting your locks into a thousand pieces to avoid false sharing. For many use cases, these designs are plenty fast enough; it turns out a lot of time and effort has gone into making channels and locks perform well! And if the simple approach is fast enough for your use case, why introduce more complex and error-prone code?

If your benchmarks indicate a performance problem, then figure out exactly which part of your system scales poorly. Focus on fixing that bottleneck in isolation where you can, and try to do so with small adjustments where possible. Maybe it's enough to split a lock in two rather than move to a concurrent hash table, or to introduce another thread and a channel rather than implement a lock-free work stealing queue. If so, do that.

Even when you do have to work directly with atomics and the like, keep things simple until there's a proven need to optimize—use `Ordering::SeqCst` and `compare_exchange` at first, and then iterate if you find concrete evidence that those are becoming bottlenecks that must be taken care of.

Write Stress Tests

As the author, you have a lot of insight into where bugs in your code may hide, without necessarily knowing what those bugs are (yet, anyway). Writing stress tests is a good way to shake out some of the hidden bugs. Stress tests don't necessarily perform a complex sequence of steps but instead have lots of threads doing relatively simple operations in parallel.

For example, if you were writing a concurrent hash map, one stress test might be to have N threads insert or update keys and M threads read keys in such a way that those $M+N$ threads are likely to often choose the same keys. Such a test doesn't test for a particular outcome or value but instead tries to trigger many possible interleavings of operations in the hopes that buggy interleavings might reveal themselves.

Stress tests resemble fuzz tests in many ways; whereas fuzzing generates many random inputs to a given function, the stress test instead generates many random thread and memory access schedules. Just like fuzzers, stress tests are therefore only as good as the assertions in your code; they can't tell you about a bug that doesn't manifest in some easy-to-spot way like an assertion failure or some other kind of panic. For that reason, it's a good idea to litter your low-level concurrency code with assertions, or `debug_assert_*` if you're worried about runtime cost in particularly hot loops.

Use Concurrency Testing Tools

The primary challenge in writing concurrent code is to handle all the possible ways the execution of different threads can interleave. As we saw in the `Ordering::SeqCst` example in Listing 10-4, it's not just the thread scheduling that matters, but also which memory values are possible for a given thread to observe at any given point in time. Writing tests that execute every possible legal execution is not only tedious but also difficult—you need very low-level control over which threads execute when and what values their reads return, which the operating system likely doesn't provide.

Model Checking with Loom

Luckily, a tool already exists that can simplify this execution exploration for you in the form of the loom crate. Given the relative release cycles of this book and that of a Rust crate, I won't give any examples of how to use Loom here, as they'd likely be out of date by the time you read this book, but I will give an overview of what it does.

Loom expects you to write dedicated test cases in the form of closures that you pass into a Loom model. The model keeps track of all cross-thread interactions and tries to intelligently explore all possible iterations of those interactions by executing the test case closure multiple times. To detect and control thread interactions, Loom provides replacement types for all the types in the standard library that allow threads to coordinate with one another; that includes most types under std::sync and std::thread as well as UnsafeCell and a few others. Loom expects your application to use those replacement types whenever you run the Loom tests. The replacement types tie into the Loom executor and perform a dual function: they act as rescheduling points so that Loom can choose which operation to run next after each possible thread interaction point, and they inform Loom of new possible interleavings to consider. Essentially, Loom builds up a tree of all the possible future executions for each point at which multiple execution interleavings are possible and then tries to execute all of them, one after the other.

Loom attempts to fully explore all possible executions of the test cases you provide it with, which means it can find bugs that occur only in extremely rare executions that stress testing would not find in a hundred years. While that's great for smaller test cases, it's generally not feasible to apply that kind of rigorous testing to larger test cases that test more involved sequences of operations or require many threads to run at once. Loom would simply take too long to get decent coverage of the code. In practice, you may therefore want to tell Loom to consider only a subset of the possible executions, which Loom's documentation has more details on.

Like with stress tests, Loom can catch only bugs that manifest as panics, so that's yet another reason to spend some time placing strategic assertions in your concurrent code! In many cases, it may even be worthwhile to add additional state tracking and bookkeeping instructions to your concurrent code to give you better assertions.

Runtime Checking with ThreadSanitizer

For larger test cases, your best bet is to run the test through a couple of iterations under Google's excellent ThreadSanitizer, also known as TSan. TSan automatically augments your code by placing extra bookkeeping instructions prior to every memory access. Then, as your code runs, those bookkeeping instructions update and check a special state machine that flags any concurrent memory operations that indicate a problematic race condition. For example, if thread B writes to some atomic value X, but has not synchronized (lots of hand waving here) with the thread that wrote the previous value of X that indicates a write/write race, which is nearly always a bug.

Since TSan only observes your code running and does not execute it over and over again like Loom, it generally only adds a constant-factor overhead to the runtime of your program. While that factor can be significant (5–15 times at the time of writing), it's still small enough that you can execute even most complex test cases in a reasonable amount of time.

At the time of writing, to use TSan you need to use a nightly version of the Rust compiler and pass in the -Zsanitizer=thread command-line argument (or set it in RUSTFLAGS), though hopefully in time this will be a standard supported option. Other sanitizers are also available that check things like out-of-bounds memory accesses, use-after-free, memory leaks, and reads of uninitialized memory, and you may want to run your concurrent test suite through those too!

HEISENBUGS

Heisenbugs are bugs that seem to disappear when you try to study them. This happens quite frequently when trying to debug highly concurrent code; the additional instrumentation to debug the problem changes the relative timing of concurrent events and might cause the execution interleaving that triggered the bug to no longer happen.

A particularly common cause of disappearing concurrency bugs is using print statements, which is by far one of the most common debugging techniques. There are two reasons why print statements have such an outsized effect on concurrency bugs. The first, and perhaps most obvious, is that relatively speaking, printing something to the user's terminal (or wherever standard output points) takes quite a long time, especially if your program is producing a lot of output. Writing to the terminal requires, at the very least, a round-trip to the operating system kernel to perform the write, but the write may also have to wait for the terminal itself to read from the process's output into its own buffers. All that extra time might so much delay the operation that previously raced with an operation in some other thread that the race condition disappears.

The second reason why print statements disturb concurrent execution patterns is that writing to standard output is (generally) guarded by a lock. If you look inside the Stdout type in the standard library, you'll see that it holds a Mutex that guards access to the output stream. It does this so that the output isn't garbled too badly if multiple threads try to write at the same time—without a lock, a given line might have characters interspersed from multiple thread writes, but with the lock the threads will take turns writing instead. Unfortunately, acquiring the output lock, is another thread synchronization point, and one that every printing thread is involved in. This means that if your code was previously broken due to missing synchronization between two threads, or just because a particular race between two threads was possible, adding print statements might fix that bug as a side effect!

In general, when you spot what seems like a Heisenbug, try to find other ways to narrow down the problem. That might involve using Loom or TSan,

(continued)

using gdb or lldb, or using a per-thread in-memory log that you print only at the end. Many logging frameworks also work hard to avoid synchronization points on the critical path of issuing log events, so switching to one of those might make your life easier. As an added bonus, good logging that you leave behind after fixing a particular bug might come in handy later. Personally I'm a big fan of the tracing crate, but there are many good options out there.

Summary

In this chapter, we first covered common correctness and performance pitfalls in concurrent Rust, and some of the high-level concurrency patterns that successful concurrent applications tend to use to work around them. We also explored how asynchronous Rust enables concurrency without parallelism, and how to explicitly introduce parallelism in asynchronous Rust code. We then dove deeper into Rust's many different lower-level concurrency primitives, including how they work, how they differ, and what they're all for. Finally, we explored techniques for writing better concurrent code and looked at tools like Loom and TSan that can help you vet that code. In the next chapter we'll continue our journey through the lower levels of Rust by digging into foreign function interfaces, which allow Rust code to link directly against code written in other languages.

11

FOREIGN FUNCTION INTERFACES

Not all code is written in Rust. It's shocking, I know. Every so often, you'll need to interact with code written in other languages, either by calling into such code from Rust or by allowing that code to call your Rust code. You can achieve this through *foreign function interfaces (FFI)*.

In this chapter we'll first look at the primary mechanism Rust provides for FFI: the extern keyword. We'll see how to use extern both to expose Rust functions and statics to other languages and to give Rust access to functions and static variables provided from outside the Rust bubble. Then, we'll walk through how to align Rust types with types defined in other languages and explore some of the intricacies of allowing data to flow across the FFI boundary. Finally, we'll talk about some of the tools you'll likely want to use if you're doing any nontrivial amount of FFI.

NOTE *While I often refer to FFI as being about crossing the boundary between one language and another, FFI can also occur entirely inside Rust-land. If one Rust program shares memory with another Rust program but the two aren't compiled together—say, if you're using a dynamically linked library in your Rust program that happens to be written in Rust, but you just have the C-compatible .so file—the same complications arise.*

Crossing Boundaries with extern

FFI is, ultimately, all about accessing bytes that originate somewhere outside your application's Rust code. For that, Rust provides two primary building blocks: *symbols*, which are names assigned to particular addresses in a given segment of your binary that allow you to share memory (be it for data or code) between the external origin and your Rust code, and *calling conventions* that provide a common understanding of how to call functions stored in such shared memory. We'll look at each of these in turn.

Symbols

Any binary artifact that the compiler produces from your code is filled with symbols—every function or static variable you define has a symbol that points to its location in the compiled binary. Generic functions may even have multiple symbols, one for each monomorphization of the function the compiler generates!

Normally, you don't have to think about symbols—they're used internally by the compiler to pass around the final address of a function or static variable in your binary. This is how the compiler knows what location in memory each function call should target when it generates the final machine code, or where to read from if your code accesses a static variable. Since you don't usually refer to symbols directly in your code, the compiler defaults to choosing semirandom names for them—you may have two functions called foo in different parts of your code, but the compiler will generate distinct symbols from them so that there's no confusion.

However, using random names for symbols won't work when you want to call a function or access a static variable that isn't compiled at the same time, such as code that's written in a different language and thus compiled by a different compiler. You can't tell Rust about a static variable defined in C if the symbol for that variable has a semirandom name that keeps changing. Conversely, you can't tell Python's FFI interface about a Rust function if you can't produce a stable name for it.

To use a symbol with an external origin, we also need some way to tell Rust about a variable or function in such a manner that the compiler will look for that same symbol defined elsewhere rather than defining its own (we'll talk about how that search happens later). Otherwise, we would just end up with two identical symbols for that function or static variable, and no sharing would take place. In fact, in all likelihood, compilation would fail since any code that referred to that symbol wouldn't know which definition (that is, which address) to use for it!

A quick note about terminology: a symbol can be declared multiple times but defined only once. Every declaration of a symbol will link to the same single definition for that symbol at linking time. If no definition for a declaration is found, or if there are multiple definitions, the linker will complain.

An Aside on Compilation and Linking

Compiler crash course time! Having a rough idea of the complicated process of turning code into a runnable binary will help you understand FFI better. You see, the compiler isn't one monolithic program but is (typically) broken down into a handful of smaller programs that each perform distinct tasks and run one after the other. At a high level, there are three distinct phases to compilation—*compilation, code generation*, and *linking*—handled by three different components.

The first phase is performed by what most people tend to think of as "the compiler"; it deals with type checking, borrow checking, monomorphization, and other features we associate with a given programming language. This phase generates no machine code but rather a low-level representation of the code that uses heavily annotated abstract machine operations. That low-level representation is then passed to the code generation tool, which is what produces machine code that can actually run on a given CPU.

These two operations, taken together, do not have to be run in a single big pass over the whole codebase all at once. Instead, the codebase can be sliced into smaller chunks that are then run through compilation concurrently. For example, Rust generally compiles different crates independently and in parallel as long as there isn't a dependency between them. It can also invoke the code generation tool for independent crates separately to process them in parallel. Rust can often even compile multiple smaller slices of a single crate separately!

Once the machine code for every piece of the application has been generated, those pieces can then be wired together. This is done in the linking phase by, unsurprisingly, the linker. The linker's primary job is to take all the binary artifacts, called *object files*, produced by code generation, stitch them together into a single file, and then replace every reference to a symbol with the final memory address of that symbol. This is how you can define a function in one crate and call it from another but still compile the two crates separately.

The linker is what enables FFI to work. It doesn't care how each of the input object files were constructed; it just dutifully links together all the object files and then resolves any shared symbols. One object file may originally have been Rust code, one originally C code, and one may be a binary blob downloaded from the internet; as long as they all use the same symbol names, the linker will make sure that the resulting machine code uses the correct cross-referenced addresses for any shared symbols.

A symbol can be linked either *statically* or *dynamically*. Static linking is the simplest, as each reference to a symbol is simply replaced with the address of that symbol's definition. Dynamic linking, on the other hand, ties each reference to a symbol to a bit of generated code that tries to find the symbol's definition when the program *runs*. We'll talk more about these linking modes a little later. Rust generally defaults to static linking for Rust code, and dynamic linking for FFI.

Using extern

The extern keyword is the mechanism that allows us to declare a symbol as residing within a foreign interface. Specifically, it declares the existence of a symbol that's defined elsewhere. In Listing 11-1 we define a static variable called RS_DEBUG in Rust that we make available to other code via FFI. We also declare a static variable called FOREIGN_DEBUG whose definition is unspecified but will be resolved at linking time.

```
#[no_mangle]
pub static RS_DEBUG: bool = true;

extern {
    static FOREIGN_DEBUG: bool;
}
```

Listing 11-1: Exposing a Rust static variable, and accessing one declared elsewhere, through FFI

The #[no_mangle] attribute ensures that RS_DEBUG retains that name during compilation rather than having the compiler assign it another symbol name to, for example, distinguish it from another (non-FFI) RS_DEBUG static variable elsewhere in the program. The variable is also declared as pub since it's a part of the crate's public API, though that annotation isn't strictly necessary on items marked #[no_mangle]. Note that we don't use extern for RS_DEBUG, since it's defined here. It will still be accessible to link against from other languages.

The extern block surrounding the FOREIGN_DEBUG static variable denotes that this declaration refers to a location that Rust will learn at linking time based on where the definition of the same symbol is located. Since it's defined elsewhere, we don't give it an initialization value, just a type, which should match the type used at the definition site. Because Rust doesn't know anything about the code that defines the static variable, and thus can't check that you've declared the correct type for the symbol, FOREIGN_DEBUG can be accessed only inside an unsafe block.

NOTE   *Static variables in Rust aren't mutable by default, regardless of whether they're in an extern block. These variables are always available from any thread, so mutable access would pose a data race risk. You can declare a static as mut, but if you do, it becomes unsafe to access.*

The procedure to declare FFI functions is very similar. In Listing 11-2, we make hello_rust accessible to non-Rust code and pull in the external hello_foreign function.

```
#[no_mangle]
pub extern fn hello_rust(i: i32) { ... }

extern {
    fn hello_foreign(i: i32);
}
```

Listing 11-2: Exposing a Rust function, and accessing one defined elsewhere, through FFI

The building blocks are all the same as in Listing 11-1 with the exception that the Rust function is declared using extern fn, which we'll explore in the next section.

If there are multiple definitions of a given extern symbol like FOREIGN_DEBUG or hello_foreign, you can explicitly specify which library the symbol should link against using the #[link] attribute. If you don't, the linker will give you an error saying that it's found multiple definitions for the symbol in question. For example, if you prefix an extern block with #[link(name = "crypto")], you're telling the linker to resolve any symbols (whether statics or functions) against a linked library named "crypto." You can also rename an external static or function in your Rust code by annotating its declaration with #[link_name = " *<actual_symbol_name>* "], and then the item links to whatever name you wish. Similarly, you can rename a Rust item for export using #[export_name = " *<export_symbol_name>* "].

Link Kinds

#[link] also accepts the argument kind, which dictates how the items in the block should be linked. The argument defaults to "dylib", which signifies C-compatible dynamic linking. The alternative kind value is "static", which indicates that the items in the block should be linked fully at compile time (that is, statically). This essentially means that the external code is wired directly into the binary produced by the compiler , and thus doesn't need to exist at runtime. There are a few other kinds as well, but they are much less common and outside the scope of this book.

There are several trade-offs between static and dynamic linking, but the main considerations are security, binary size, and distribution. First, dynamic linking tends to be more secure because it makes it easier to upgrade libraries independently. Dynamic linking allows whoever deploys a binary that contains your code to upgrade libraries your code links against without having to recompile your code. If, say, libcrypto gets a security update, the user can update the crypto library on the host and restart the binary, and the updated library code will be used automatically. With static compilation, the library's code is hardwired into the binary, so the user would have to recompile your code against an upgraded version of the library to get the update.

Dynamic linking also tends to produce smaller binaries. Since static compilation includes any linked code into the final binary output, and any code that code in turn pulls in, it produces larger binaries. With dynamic linking, each external item includes just a small bit of wrapper code that loads the indicated library at runtime and then forwards the access.

So far, static linking may not seem very attractive, but it has one big advantage over dynamic linking: ease of distribution. With dynamic linking, anyone who wants to run a binary that includes your code must *also* have any libraries your code links against. Not only that, but they must make sure the version of each such library they have is compatible with what your code expects. This may be fine for libraries like glibc or OpenSSL that are available on most systems, but it poses a problem for more obscure libraries. The user then needs to be aware that they should install that library and must hunt for it in order to run your code! With static linking, the library's code is embedded directly into the binary output, so the user doesn't need to install it themselves.

Ultimately, there isn't a *right* choice between static and dynamic linking. Dynamic linking is usually a good default, but static compilation may be a better option for particularly constrained deployment environments or for very small or niche library dependencies. Use your best judgment!

Calling Conventions

Symbols dictate *where* a given function or variable is defined, but that's not enough to allow function calls across FFI boundaries. To call a foreign function in any language, the compiler also needs to know its *calling convention*, which dictates the assembly code to use to invoke the function. We won't get into the actual technical details of each calling convention here, but as a general overview, the convention dictates:

- How the stack frame for the call is set up
- How arguments are passed (whether on the stack or in registers, in order or in reverse)
- How the function is told where to jump back to when it returns
- How various CPU states, like registers, are restored in the caller after the function completes

Rust has its own unique calling convention that isn't standardized and is allowed to be changed by the compiler over time. This works fine as long as all function definitions and calls are compiled by the same Rust compiler, but it is problematic if you want interoperability with external code because that external code doesn't know about the Rust calling convention.

Every Rust function is implicitly declared with extern "Rust" if you don't declare anything else. Using extern on its own, as in Listing 11-2, is shorthand for extern "C", which means "use the standard C calling convention." The shorthand is there because the C calling convention is what you want in nearly every case of FFI.

NOTE *Unwinding generally works only with regular Rust functions. If you unwind across the end of a Rust function that isn't extern "Rust", your program will abort. Unwinding across the FFI boundary into external code is undefined behavior. With RFC 2945, Rust gained a new extern declaration, extern "C-unwind"; this permits unwinding across FFI boundaries in particular situations, but if you wish to use it you should read the RFC carefully.*

Rust also supports a number of other calling conventions that you supply as a string following the extern keyword (in both fn and block context). For example, extern "system" says to use the calling convention of the operating system's standard library interface, which at the time of writing is the same as "C" everywhere except on Win32, which uses the "stdcall" calling convention. In general, you'll rarely need to supply a calling convention explicitly unless you're working with particularly platform-specific or highly optimized external interfaces, so just extern (which is extern "C") will be fine.

NOTE *A function's calling convention is part of its type. That is, the type extern "C" fn() is not the same as fn() (or extern "Rust" fn()), which is different again from extern "system" fn().*

OTHER BINARY ARTIFACTS

Normally, you compile Rust code only to run its tests or build a binary that you're then going to distribute or run. Unlike in many other languages, you don't generally compile a Rust library to distribute it to others—if you run a command like cargo publish, it just wraps up your crate's source code and uploads it to *crates.io*. This is mostly because it is difficult to distribute generic code as anything but source code. Since the compiler monomorphizes each generic function to the provided type arguments, and those types may be defined in the caller's crate, the compiler must have access to the function's *generic* form, which means no optimized machine code!

Technically speaking, Rust does compile binary library artifacts, called *rlibs*, of each dependency that it combines in the end. These rlibs include the information necessary to resolve generic types, but they are specific to the exact compiler used and can't generally be distributed in any meaningful way.

So what do you do if you want to write a library in Rust that you then want to interface with from another programming language? The solution is to produce C-compatible library files in the form of dynamically linked libraries (*.so* files on Unix, *.dylib* files on macOS, and *.dll* files on Windows) and statically linked libraries (*.a* files on Unix/macOS and *.lib* files on Windows). Those files look like files produced by C code, so they can also be used by other languages that know how to interact with C.

To produce these C-compatible binary artifacts, you set the crate-type field of the [lib] section of your *Cargo.toml* file. The field takes an array of values, which would normally just be "lib" to indicate a standard Rust library (an rlib). Cargo applies some heuristics that will set this value automatically if your crate is clearly not a library (for example, if it's a procedural macro), but best practice is to set this value explicitly if you're producing anything but a good ol' Rust library.

There are a number of different crate types, but the relevant ones here are "cdylib" and "staticlib", which produce C-compatible library files that are dynamically and statically linked, respectively. Keep in mind that when you

(continued)

produce one of these artifact types, only publicly available symbols are available—that is, public and #[no_mangle] static variables and functions. Things like types and constants won't be available, even if they're marked pub, since they have no meaningful representation in a binary library file.

Types Across Language Boundaries

With FFI, type layout is crucial; if one language lays out the memory for some shared data one way but the language on the other side of the FFI boundary expects it to be laid out differently, then the two sides will interpret the data inconsistently. In this section, we'll look at how to make types match up over FFI, and other aspects of types to be aware of when you cross the boundaries between languages.

Type Matching

Types aren't shared across the FFI boundary. When you declare a type in Rust, that type information is lost entirely upon compilation. All that's communicated to the other side is the bits that make up values of that type. You therefore need to declare the type for those bits on both sides of the boundary. When you declare the Rust version of the type, you first must make sure the primitives contained within the type match up. For example, if C is used on the other side of the boundary, and the C type uses an int, the Rust code had better use the exact Rust equivalent: an i32. To take some of the guesswork out of that process, for interfaces that use C-like types the Rust standard library provides you with the correct C types in the std::os::raw module, which defines type c_int = i32, type c_char = i8/u8 depending on whether char is signed, type c_long = i32/i64 depending on the target pointer width, and so on.

NOTE *Take particular note of quirky integer types in C like __be32. These often do not translate directly to Rust types and may be best left as something like [u8; 4]. For example, __be32 is always encoded as big-endian, whereas Rust's i32 uses the endianness of the current platform.*

With more complex types like vectors and strings, you usually need to do the mapping manually. For example, since C tends to represent a string as a sequence of bytes terminated with a 0 byte, rather than a UTF-8–encoded string with the length stored separately, you cannot generally use Rust's string types over FFI. Instead, assuming the other side uses a C-style string representation, you should use the std::ffi::CStr and std::ffi::CString types for borrowed and owned strings, respectively. For vectors, you'll likely want to use a raw pointer to the first element and then pass the length separately—the Vec::into_raw_parts method may come in handy for that.

For types that contain other types, such as structs and unions, you also need to deal with layout and alignment. As we discussed in Chapter 2, Rust lays out types in an undefined way by default, so at the very least you will want to use #[repr(C)] to ensure that the type has a deterministic layout and alignment that mirrors what's (likely and hopefully) used across the FFI boundary. If the interface also specifies other configurations for the type, such as manually setting its alignment or removing padding, you'll need to adjust your #[repr] accordingly.

A Rust enum has multiple possible C-style representations depending on whether the enum contains data or not. Consider an enum without data, like this:

```
enum Foo { Bar, Baz }
```

With #[repr(C)], the type Foo is encoded using just a single integer of the same size that a C compiler would choose for an enum with the same number of variants. The first variant has the value 0, the second the value 1, and so on. You can also manually assign values to each variant, as shown in Listing 11-3.

```
#[repr(C)]
enum Foo {
    Bar = 1,
    Baz = 2,
}
```

Listing 11-3: Defining explicit variant values for a dataless enum

NOTE *Technically, the specification says that the first variant's value is 0 and every subsequent variant's value is one greater than that of the previous one. This makes a difference if you manually set the value for some variants but not others—those you do not set will continue from the last one you did set.*

You should be careful about mapping enum-like types in C to Rust this way, however, as only the values for defined variants are valid for an instance of the enum type. This tends to get you into trouble with C-style enumerations that often function more like bitsets, where variants can be bitwise ORed together to produce a value that encapsulates multiple variants at once. In the example from Listing 11-3, for instance, a value of 3 produced by taking Bar | Baz would not be valid for Foo in Rust! If you need to model a C API that uses an enumeration for a set of bitflags that can be set and unset individually, consider using a newtype wrapper around an integer type, with associated constants for each variant and implementations of the various Bit* traits for improved ergonomics. Or use the bitflags crate.

NOTE *For fieldless enums, you can also pass a numeric type to #[repr] to use a different type than isize for the discriminator. For example, #[repr(u8)] will encode the discriminator using a single unsigned byte. For a data-carrying enum, you can pass #[repr(C, u8)] to get the same effect.*

On an enum that contains data, the #[repr(C)] attribute causes the enum to be represented using a *tagged union*. That is, it is represented in memory by a #[repr(C)] struct with two fields, where the first is the discriminator as it would be encoded if none of the variants had fields, and the second is a union of the data structures for each variant. For a concrete example, consider the enum and associated representation in Listing 11-4.

```
#[repr(C)]
enum Foo {
    Bar(i32),
    Baz { a: bool, b: f64 }
}
// is represented as
#[repr(C)]
enum FooTag { Bar, Baz }
#[repr(C)]
struct FooBar(i32);
#[repr(C)]
struct FooBaz{ a: bool, b: f64 }
#[repr(C)]
union FooData {
  bar: FooBar,
  baz: FooBaz,
}
#[repr(C)]
struct Foo {
    tag: FooTag,
    data: FooData
}
```

Listing 11-4: Rust enums with #[repr(C)] are represented as tagged unions.

THE NICHE OPTIMIZATION IN FFI

In Chapter 9 we talked about the niche optimization, where the Rust compiler uses invalid bit patterns to represent enum variants that hold no data. The fact that this optimization is guaranteed leads to an interesting interaction with FFI. Specifically, it means that nullable pointers can always be represented in FFI types using an Option-wrapped pointer type. For example, a nullable function pointer can be represented as Option<extern fn(...)>, and a nullable data pointer can be represented as Option<*mut T>. These will transparently do the right thing if an all-zero bit pattern value is provided, and will represent it as None in Rust.

Allocations

When you allocate memory, that allocation belongs to its allocator and can be freed only by that same allocator. This is the case if you use multiple allocators within Rust and also if you are allocating memory both in Rust and with some allocator on the other side of the FFI boundary. You're free to send pointers across the boundary and access that memory to your heart's content, but when it comes to releasing the memory again, it needs to be returned to the appropriate allocator.

Most FFI interfaces will have one of two configurations for handling allocation: either the caller provides data pointers to chunks of memory or the interface exposes dedicated freeing methods to which any allocated resources should be returned when they are no longer needed. Listing 11-5 shows an example of Rust declarations of some signatures from the OpenSSL library that use implementation-managed memory.

```
// One function allocates memory for a new object.
extern fn ECDSA_SIG_new() -> *mut ECDSA_SIG;

// And another accepts a pointer created by new
// and deallocates it when the caller is done with it.
extern fn ECDSA_SIG_free(sig: *mut ECDSA_SIG);
```

Listing 11-5: An implementation-managed memory interface

The functions ECDSA_SIG_new and ECDSA_SIG_free form a pair, where the caller is expected to call the new function, use the returned pointer for as long as it needs (likely by passing it to other functions in turn), and then finally pass the pointer to the free function once it's done with the referenced resource. Presumably, the implementation allocates memory in the new function and deallocates it in the free function. If these functions were defined in Rust, the new function would likely use Box::new, and the free function would invoke Box::from_raw and then drop the value to run its destructor.

Listing 11-6 shows an example of caller-managed memory.

```
// An example of caller-managed memory.
// The caller provides a pointer to a chunk of memory,
// which the implementation then uses to instantiate its own types.
// No free function is provided, as that happens in the caller.
extern fn BIO_new_mem_buf(buf: *const c_void, len: c_int) -> *mut BIO
```

Listing 11-6: A caller-managed memory interface

Here, the BIO_new_mem_buf function instead has the caller supply the backing memory. The caller can choose to allocate memory on the heap, or use whatever other mechanism it deems fit for obtaining the required memory, and then passes it to the library. The onus is then on the caller to ensure that the memory is later deallocated, but only once it is no longer needed by the FFI implementation!

You can use either of these approaches in your FFI APIs or even mix and match them if you wish. As a general rule of thumb, allow the caller to pass in memory when doing so is feasible, since it gives the caller more freedom to manage memory as it deems appropriate. For example, the caller may be using a highly specialized allocator on some custom operating system, and may not want to be forced to use the standard allocator your implementation would use. If the caller can pass in the memory, it might even avoid allocations entirely if it can instead use stack memory or reuse already allocated memory. However, keep in mind that the ergonomics of a caller-managed interface are often more convoluted, since the caller must now do all the work to figure out how much memory to allocate and then set that up before it can call into your library.

In some instances, it may even be impossible for the caller to know ahead of time how much memory to allocate—for example, if your library's types are opaque (and thus not known to the caller) or can change over time, the caller won't be able to predict the size of the allocation. Similarly, if your code has to allocate more memory while it is running, such as if you're building a graph on the fly, the amount of memory needed may vary dynamically at runtime. In such cases, you will have to use implementation-managed memory.

When you're forced to make a trade-off, go with caller-allocated memory for anything that is either *large* or *frequent*. In those cases the caller is likely to care the most about controlling the allocations itself. For anything else, it's probably okay for your code to allocate and then expose destructor functions for each relevant type.

Callbacks

You can pass function pointers across the FFI boundary and call the referenced function through those pointers as long as the function pointer's type has an extern annotation that matches the function's calling convention. That is, you can define an extern "C" fn(c_int) -> c_int in Rust and then pass a reference to that function to C code as a callback that the C code will eventually invoke.

You do need to be careful using callbacks around panics, as having a panic unwind past the end of a function that is anything but extern "Rust" is undefined behavior. The Rust compiler will currently automatically abort if it detects such a panic, but that may not always be the behavior you want. Instead, you may want to use std::panic::catch_unwind to detect the panic in any function marked extern, and then translate the panic into an error that is FFI-compatible.

Safety

When you write Rust FFI bindings, most of the code that actually interfaces with the FFI will be unsafe and will mainly revolve around raw pointers. However, your goal should be to ultimately present a *safe* Rust interface on top of the FFI. Doing so mainly comes down to reading carefully through the invariants of the unsafe interface you are wrapping and then ensuring you uphold them all through the Rust type system in the safe interface. The three most important elements of safely encapsulating a foreign interface are capturing & versus &mut accurately, implementing Send and Sync appropriately, and ensuring that pointers cannot be accidentally confused. I'll go over how to enforce each of these next.

References and Lifetimes

If there's a chance external code will modify data behind a given pointer, make sure that the safe Rust interface has an exclusive reference to the relevant data by taking &mut. Otherwise a user of your safe wrapper might accidentally read from memory that the external code is simultaneously modifying, and all hell will break loose!

You'll also want to make good use of Rust lifetimes to ensure that all pointers live for as long as the FFI requires. For example, imagine an external interface that lets you create a Context and then lets you create a Device from that Context with the requirement that the Context remain valid for as long as the Device lives. In that case, any safe wrapper for the interface should enforce that requirement in the type system by having Device hold a lifetime associated with the borrow of Context that the Device was created from.

Send and Sync

Do not implement Send and Sync for types from an external library unless that library explicitly documents that those types are thread-safe! It is the safe Rust wrapper's job to ensure that safe Rust code *cannot* violate the invariants of the external code and thus trigger undefined behavior.

Sometimes, you may even want to introduce dummy types to enforce external invariants. For example, say you have an event loop library with the interface given in Listing 11-7.

```
extern fn start_main_loop();
extern fn next_event() -> *mut Event;
```

Listing 11-7: A library that expects single-threaded use

Now suppose that the documentation for the external library states that next_event may be called only by the same thread that called start_main_loop. However, here we have no type that we can avoid implementing Send for! Instead, we can take a page out of Chapter 3 and introduce additional marker state to enforce the invariant, as shown in Listing 11-8.

```
pub struct EventLoop(std::marker::PhantomData<*const ()>);
pub fn start() -> EventLoop {
    unsafe { ffi::start_main_loop() };
    EventLoop(std::marker::PhantomData)
}
impl EventLoop {
    pub fn next_event(&self) -> Option<Event> {
        let e = unsafe { ffi::next_event() };
        // ...
    }
}
```

Listing 11-8: Enforcing an FFI invariant by introducing auxiliary types

The empty type EventLoop doesn't actually connect with anything in the underlying external interface but rather enforces the contract that you call next_event only after calling start_main_loop, and only on the same thread. You enforce the "same thread" part by making EventLoop neither Send nor Sync, by having it hold a phantom raw pointer (which itself is neither Send nor Sync).

Using PhantomData<*const ()> to "undo" the Send and Sync auto-traits as we do here is a bit ugly and indirect. Rust does have an unstable compiler feature that enables negative trait implementations like impl !Send for EventLoop {}, but it's surprisingly difficult to get its implementation right, and it likely won't stabilize for some time.

You may have noticed that nothing prevents the caller from invoking start_main_loop multiple times, either from the same thread or from another thread. How you'd handle that would depend on the semantics of the library in question, so I'll leave it to you as an exercise.

Pointer Confusion

In many FFI APIs, you don't necessarily want the caller to know the internal representation for each and every chunk of memory you give it pointers to. The type might have internal state that the caller shouldn't fiddle with, or the state might be difficult to express in a cross-language-compatible way. For these kinds of situations, C-style APIs usually expose *void pointers*, written out as the C type void*, which is equivalent to *mut std::ffi::c_void in Rust. A type-erased pointer like this is, effectively, *just* a pointer, and does not convey anything about the thing it points to. For that reason, these kinds of pointers are often referred to as *opaque*.

Opaque pointers effectively serve the role of visibility modifiers for types across FFI boundaries—since the method signature does not say what's being pointed to, the caller has no option but to pass around the pointer as is and use any available FFI methods to provide visibility into the referenced data. Unfortunately, since one *mut c_void is indistinguishable from another, there's nothing stopping a user from taking an opaque pointer as is returned from one FFI method and supplying it to a method that expects a pointer to a *different* opaque type.

We can do better than this in Rust. To mitigate this kind of pointer type confusion, we can avoid using *mut c_void directly for opaque pointers in FFI, even if the actual interface calls for a void*, and instead construct different empty types for each distinct opaque type. For example, in Listing 11-9 I use two distinct opaque pointer types that cannot be confused.

```
#[non_exhaustive] #[repr(transparent)] pub struct Foo(c_void);
#[non_exhaustive] #[repr(transparent)] pub struct Bar(c_void);
extern {
    pub fn foo() -> *mut Foo;
    pub fn take_foo(arg: *mut Foo);
    pub fn take_bar(arg: *mut Bar);
}
```

Listing 11-9: Opaque pointer types that cannot be confused

Since Foo and Bar are both zero-sized types, they can be used in place of () in the extern method signatures. Even better, since they are now distinct types, Rust won't let you use one where the other is required, so it's now impossible to call take_bar with a pointer you got back from foo. Adding the #[non_exhaustive] annotation ensures that the Foo and Bar types cannot be constructed outside of this crate.

bindgen and Build Scripts

Mapping out the Rust types and externs for a larger external library can be quite a chore. Big libraries tend to have a large enough number of type and method signatures to match up that writing out all the Rust equivalents is time-consuming. They also have enough corner cases and C oddities that some patterns are bound to require more careful thought to translate.

Luckily, the Rust community has developed a tool called bindgen that significantly simplifies this process as long as you have C header files available for the library you want to interface with. bindgen essentially encodes all the rules and best practices we've discussed in this chapter, plus a number of others, and wraps them up in a configurable code generator that takes in C header files and spits out appropriate Rust equivalents.

bindgen provides a stand-alone binary that generates the Rust code for C headers once, which is convenient when you want to check in the bindings. This process allows you to hand-tune the generated bindings, should that be necessary. If, on the other hand, you want to generate the bindings automatically on every build and just include the C header files in your source code, bindgen also ships as a library that you can invoke in a custom *build script* for your package.

NOTE *If you check in the bindings directly, keep in mind that they will be correct only on the platform they were generated for. Generating the bindings in a build script will generate them specifically for the current target platform, which is less likely to cause platform-related layout inconsistencies.*

You declare a build script by adding build = " <some-file.rs> " to the [package] section of your *Cargo.toml*. This tells Cargo that, before compiling your crate, it should compile <some-file.rs> as a stand-alone Rust program and run it; only then should it compile the source code of your crate. The build script also gets its own dependencies, which you declare in the [build-dependencies] section of your *Cargo.toml*.

NOTE *If you name your build script* build.rs, *you don't need to declare it in your* Cargo.toml.

Build scripts come in very handy with FFI—they can compile a bundled C library from source, dynamically discover and declare additional build flags to be passed to the compiler, declare additional files that Cargo should check for changes for the purposes of recompilation, and, you guessed it, generate additional source files on the fly!

Though build scripts are very versatile, beware of making them too aware of the environment they run in. While you can use a build script to detect if the Rust compiler version is a prime or if it's going to rain in Istanbul tomorrow, making your compilation dependent on such conditions may make builds fail unexpectedly for other developers, which leads to a poor development experience.

The build script can write files to a special directory supplied through the OUT_DIR environment variable. The same directory and environment variable are also accessible in the Rust source code at compile time so that it can pick up files generated by the build script. To generate and use Rust types from a C header, you first have your build script use the library version of bindgen to read in a *.h* file and turn it into a file called, say, *bindings.rs* inside OUT_DIR. You then add the following line to any Rust file in your crate to include *bindings.rs* at compilation time:

```
include!(concat!(env!("OUT_DIR"), "/bindings.rs"));
```

Since the code in *bindings.rs* is autogenerated, it's generally best practice to place the bindings in their own crate and give the crate the same name as the library the bindings are for, with the suffix -sys (for example, openssl-sys). If you don't follow this practice, releasing new versions of your library will be much more painful, as it is illegal for two crates that link against the same external library through the links key in *Cargo.toml* to coexist in a given build. You would essentially have to upgrade the entire ecosystem to the new major version of your library all at once. Separating just the bindings into their own crate allows you to issue new major versions of the wrapper crate that can be adopted incrementally. The separation also allows you to cut a breaking release of the crate with those bindings if the Rust bindings change—say, if the header files themselves are upgraded or a bindgen upgrade causes the generated Rust code to change slightly—without *also* having to cut a breaking release of the crate that safely wraps the FFI bindings.

NOTE *Remember that if you include any of the types from the -sys crate in the public interface of your main library crate, changing the dependency on the -sys crate to a new major version still constitutes a breaking change for your main library!*

If your crate instead produces a library file that you intend others to use through FFI, you should also publish a C header file for its interface to make it easier to generate native bindings to your library from other languages. However, that C header file then needs to be kept up to date as your crate changes, which can become cumbersome as your library grows in size. Fortunately, the Rust community has also developed a tool to automate this task: cbindgen. Like bindgen, cbindgen is a build tool, and it also comes as both a binary and a library for use in build scripts. Instead of taking in a C header file and producing Rust, it takes Rust in and produces a C header file. Since the C header file represents the main computer-readable description of your crate's FFI, I recommend manually looking it over to make sure the autogenerated C code isn't too unwieldy, though in general cbindgen tends to produce fairly reasonable code. If it doesn't, file a bug!

C++

I've mainly focused on C in this chapter as it's the language most commonly used to describe cross-language interfaces for libraries you can link against. Nearly every programming language provides some way to interact with C libraries, since they are so ubiquitous. While C++ feels closely related to C, and many high-profile libraries are written in C++, it's a very different beast when it comes to FFI. Generating types and signatures to match a C header is relatively straightforward, but that is not at all the case for C++. At the time of writing, bindgen has decent support for generating bindings to C++, but they are often lacking in ergonomics. For example, you generally have to manually call constructors, destructors, overloaded operators, and the like. Some C++ features like template specialization also aren't supported at all. If you do have to interface with C++, I recommend you give the cxx crate a try.

Summary

In this chapter, we've covered how to use the extern keyword to call out of Rust into external code, as well as how to use it to make Rust code accessible to external code. We've also discussed how to align Rust types with types on the other side of the FFI boundary, and some of the common pitfalls in trying to get code written in two different languages to mesh well. Finally, we talked about the bindgen and cbindgen tools, which make the experience of keeping FFI bindings up to date much more pleasant. In the next chapter, we'll look at how to use Rust in more restricted environments, like embedded devices, where the standard library may not be available and where even a simple operation like allocating memory may not be possible.

12

RUST WITHOUT THE STANDARD LIBRARY

Rust is intended to be a language for systems programming, but it isn't always clear what that really means. At the very least, a systems programming language is usually expected to allow the programmer to write programs that do not rely on the operating system and can run directly on the hardware, whether that is a thousand-core supercomputer or an embedded device with a single-core ARM processor with a clock speed of 72MHz and 256KiB of memory.

In this chapter, we'll take a look at how you can use Rust in unorthodox environments, such as those without an operating system, or those that don't even have the ability to dynamically allocate memory! Much of our discussion will focus on the #![no_std] attribute, but we'll also investigate

Rust's alloc module, the Rust runtime (yes, Rust does technically have a runtime), and some of the tricks you have to play to write up a Rust binary for use in such an environment.

Opting Out of the Standard Library

As a language, Rust consists of multiple independent pieces. First there's the compiler, which dictates the grammar of the Rust language and implements type checking, borrow checking, and the final conversion into machine-runnable code. Then there's the standard library, std, which implements all the useful common functionality that most programs need—things like file and network access, a notion of time, facilities for printing and reading user input, and so on. But std itself is also a composite, building on top of two other, more fundamental libraries called core and alloc. In fact, many of the types and functions in std are just re-exports from those two libraries.

The core library sits at the bottom of the standard library pyramid and contains any functionality that depends on nothing but the Rust language itself and the hardware the resulting program is running on—things like sorting algorithms, marker types, fundamental types such as Option and Result, low-level operations such as atomic memory access methods, and compiler hints. The core library works as if the operating system does not exist, so there is no standard input, no filesystem, and no network. Similarly, there is no memory allocator, so types like Box, Vec, and HashMap are nowhere to be seen.

Above core sits alloc, which holds all the functionality that depends on dynamic memory allocation, such as collections, smart pointers, and dynamically allocated strings (String). We'll get back to alloc in the next section.

Most of the time, because std re-exports everything in core and alloc, developers do not need to know about the differences among the three libraries. This means that even though Option technically lives in core::option::Option, you can access it through std::option::Option.

However, in an unorthodox environment, such as on an embedded device where there is no operating system, the distinction is crucial. While it's fine to use an Iterator or to sort a list of numbers, an embedded device may simply have no meaningful way to access a file (as that requires a filesystem) or print to the terminal (as that requires a terminal)—so there's no File or println!. Furthermore, the device may have so little memory that dynamic memory allocation is a luxury you can't afford, and thus anything that allocates memory on the fly is a no-go—say goodbye to Box and Vec.

Rather than force developers to carefully avoid those basic constructs in such environments, Rust provides a way to opt out of anything but the core functionality of the language: the #![no_std] attribute. This is a crate-level attribute (#!) that switches the prelude (see the box on page 213) for the crate from std::prelude to core::prelude so that you don't accidentally depend on anything outside of core that might not work in your target environment.

However, that is *all* the #![no_std] attribute does—it does not prevent you from bringing in the standard library explicitly with extern std. This may be surprising, as it means a crate marked #![no_std] may in fact not be compatible with a target environment that does not support std, but this design decision was intentional: it allows you to mark your crate as being no_std-compatible but to still use features from the standard library when certain features are enabled. For example, many crates have a feature named std that, when enabled, gives access to more sophisticated APIs and integrations with types that live in std. This allows crate authors to both supply the core implementation for constrained use cases and add bells and whistles for consumers on more standard platforms.

NOTE *Since features should be additive, prefer an std-enabling feature to an std-disabling one. Otherwise, if any crate in a consumer's dependency graph enables the no-std feature, all consumers will be given access only to the bare-bones API without std support, which may then mean that APIs they depend on aren't available, causing them to no longer compile.*

THE PRELUDE

Have you ever wondered why there are some types and traits—like Box, Iterator, Option, and Clone—that are available in every Rust file without you needing to use them? Or why you don't need to use any of the macros in the standard library (like vec![])? The reason is that every Rust module automatically imports the Rust standard prelude with an implicit use std::prelude::rust_2021::* (or similar for other editions), which brings all the exports from the crate's chosen edition's prelude into scope. The prelude modules themselves aren't special beyond this auto-inclusion—they are merely collections of pub use statements for key types, traits, and macros that the Rust developers expect to be commonly used.

Dynamic Memory Allocation

As we discussed in Chapter 1, a machine has many different regions of memory, and each one serves a distinct purpose. There's static memory for your program code and static variables, there's the stack for function-local variables and function arguments, and there's the heap for, well, everything else. The heap supports allocating variably sized regions of memory at runtime, and those allocations stick around for however long you want them to. This makes heap memory extremely versatile, and as a result, you find it used everywhere. Vec, String, Arc and Rc, and the collection types are all implemented in heap memory, which allows them to grow and shrink over time and to be returned from functions without the borrow checker complaining.

Behind the scenes, the heap is really just a huge chunk of contiguous memory that is managed by an *allocator*. It's the allocator that provides the illusion of distinct allocations in the heap, ensuring that those allocations do not overlap and that regions of memory that are no longer in use are reused. By default Rust uses the system allocator, which is generally the one dictated by the standard C library. This works well for most use cases, but if necessary, you can override which allocator Rust will use through the `GlobalAlloc` trait combined with the `#[global_allocator]` attribute, which requires an implementation of an `alloc` method for allocating a new segment of memory and `dealloc` for returning a past allocation to the allocator to reuse.

In environments without an operating system, the standard C library is also generally not available, and so neither is the standard system allocator. For that reason, `#![no_std]` also excludes all types that rely on dynamic memory allocation. But since it's entirely possible to implement a memory allocator without access to a full-blown operating system, Rust allows you to opt back into just the part of the Rust standard library that requires an allocator without opting into all of `std` through the `alloc` crate. The `alloc` crate comes with the standard Rust toolchain (just like `core` and `std`) and contains most of your favorite heap-allocation types, like `Box`, `Arc`, `String`, `Vec`, and `BTreeMap`. `HashMap` is not among them, since it relies on random number generation for its key hashing, which is an operating system facility. To use types from `alloc` in a `no_std` context, all you have to do is replace any imports of those types that previously had `use std::` with `use alloc::` instead. Do keep in mind, though, that depending on `alloc` means your `#![no_std]` crate will no longer be usable by any program that disallows dynamic memory allocation, either because it doesn't have an allocator or because it has too little memory to permit dynamic memory allocation in the first place.

NOTE *Some programming domains, like the Linux kernel, may allow dynamic memory allocation only if out-of-memory errors are handled gracefully (that is, without panicking). For such use cases, you'll want to provide* try_ *versions of any methods you expose that might allocate. The* try_ *methods should use fallible methods of any inner types (like the currently unstable* Box::try_new *or* Vec::try_reserve*) rather than ones that just panic (like* Box::new *or* Vec::reserve*) and propagate those errors out to the caller, who can then handle them appropriately.*

It might strike you as odd that it's possible to write nontrivial crates that use *only* `core`. After all, they can't use collections, the `String` type, the network, or the filesystem, and they don't even have a notion of time! The trick to core-only crates is to utilize the stack and static allocations. For example, for a heapless vector, you allocate enough memory up front—either in static memory or in a function's stack frame—for the largest number of elements you expect the vector to be able to hold, and then augment it with a `usize` that tracks how many elements it currently holds. To push to the vector, you write to the next element in the (statically sized) array and increment a variable that tracks the number of elements. If the vector's length ever reaches the static size, the next push fails. Listing 12-1 gives an example of such a heapless vector type implemented using `const` generics.

```
struct ArrayVec<T, const N: usize> {
    values: [Option<T>; N],
    len: usize,
}
impl<T, const N: usize> ArrayVec<T, N> {
    fn try_push(&mut self, t: T) -> Result<(), T> {
        if self.len == N {
            return Err(t);
        }
        self.values[self.len] = Some(t);
        self.len += 1;
        return Ok(());
    }
}
```

Listing 12-1: A heapless vector type

We make ArrayVec generic over both the type of its elements, T, and the maximum number of elements, N, and then represent the vector as an array of N *optional* Ts. This structure always stores N Option<T>, so it has a size known at compile time and can be stored on the stack, but it can still act like a vector by using runtime information to inform how we access the array.

NOTE   *We could have implemented ArrayVec using [MaybeUninit<T>; N] to avoid the overhead of the Option, but that would require using unsafe code, which isn't warranted for this example.*

The Rust Runtime

You may have heard the claim that Rust doesn't have a runtime. While that's true at a high level—it doesn't have a garbage collector, an interpreter, or a built-in user-level scheduler—it's not really true in the strictest sense. Specifically, Rust does have some special code that runs before your main function and in response to certain special conditions in your code, which really is a form of bare-bones runtime.

The Panic Handler

The first bit of such special code is Rust's *panic handler*. When Rust code panics by invoking panic! or panic_any, the panic handler dictates what happens next. When the Rust runtime is available—as is the case on most targets that supply std—the panic handler first invokes the *panic hook* set via std::panic::set_hook, which prints a message and optionally a backtrace to standard error by default. It then either unwinds the current thread's stack or aborts the process, depending on the panic setting chosen for current compilation (either through Cargo configuration or arguments passed directly to rustc).

However, not all targets provide a panic handler. For example, most embedded targets do not, as there isn't necessarily a single implementation that makes sense across all the uses for such a target. For targets that don't supply a panic handler, Rust still needs to know what to do when a panic occurs. To that end, we can use the #[panic_handler] attribute to decorate a single function in the program with the signature fn(&PanicInfo) -> !. That function is called whenever the program invokes a panic, and it is passed information about the panic in the form of a core::panic::PanicInfo. What the function does with that information is entirely unspecified, but it can never return (as indicated by the ! return type). This is important, since the Rust compiler assumes that no code that follows a panic is run.

There are many valid ways for a panic handler to avoid returning. The standard panic handler unwinds the thread's stack and then terminates the thread, but a panic handler can also halt the thread using loop {}, abort the program, or do anything else that makes sense for the target platform, even as far as resetting the device.

Program Initialization

Contrary to popular belief, the main function is not the first thing that runs in a Rust program. Instead, the main symbol in a Rust binary actually points to a function in the standard library called lang_start. That function performs the (fairly minimal) setup for the Rust runtime, including stashing the program's command-line arguments in a place where std::env::args can get to them, setting the name of the main thread, handling panics in the main function, flushing standard output on program exit, and setting up signal handlers. The lang_start function in turn calls the main function defined in your crate, which then doesn't need to think about how, for example, Windows and Linux differ in how command-line arguments are passed in.

This arrangement works well on platforms where all of that setup is sensible and supported, but it presents a problem on embedded platforms where main memory may not even be accessible when the program starts. On such platforms, you'll generally want to opt out of the Rust initialization code entirely using the #![no_main] crate-level attribute. This attribute completely omits lang_start, meaning you as the developer must figure out how the program should be started, such as by declaring a function with #[export_name = "main"] that matches the expected launch sequence for the target platform.

NOTE *On platforms that truly run no code before they jump to the defined start symbol, like most embedded devices, the initial values of static variables may not even match what's specified in the source code. In such cases, your initialization function will need to explicitly initialize the various static memory segments with the initial data values specified in your program binary.*

The Out-of-Memory Handler

If you write a program that wishes to use alloc but is built for a platform that does not supply an allocator, you must dictate which allocator to use using the #[global_allocator] attribute mentioned earlier in the chapter. But you also have to specify what happens if that global allocator fails to allocate memory. Specifically, you need to define an *out-of-memory handler* to say what should happen if an infallible operation like Vec::push needs to allocate more memory, but the allocator cannot supply it.

The default behavior of the out-of-memory handler on std-enabled platforms is to print an error message to standard error and then abort the process. However, on a platform that, for example, doesn't have standard error, that obviously won't work. At the time of writing, on such platforms your program must explicitly define an out-of-memory handler using the unstable attribute #[lang = "oom"]. Keep in mind that the handler should almost certainly prevent future execution, as otherwise the code that tried to allocate will continue executing without knowing that it did not receive the memory it asked for!

NOTE *By the time you read this, the out-of-memory handler may already have been stabilized under a permanent name (#[alloc_error_handler], most likely). Work is also under-way to give the default std out-of-memory handler the same kind of "hook" function-ality as Rust's panic handler, so that code can change the out-of-memory behavior on the fly through a method like set_alloc_error_hook.*

Low-Level Memory Accesses

In Chapter 10, we discussed the fact that the compiler is given a fair amount of leeway in how it turns your program statements into machine instruc-tions, and that the CPU is allowed some wiggle room to execute instruc-tions out of order. Normally, the shortcuts and optimizations that the compiler and CPU can take advantage of are invisible to the semantics of the program—you can't generally tell whether, say, two reads have been reordered relative to each other or whether two reads from the same mem-ory location actually result in two CPU load instructions. This is by design. The language and hardware designers carefully specified what semantics programmers commonly expect from their code when it runs so that your code generally does what you expect it to.

However, no_std programming sometimes takes you beyond the usual border of "invisible optimizations." In particular, you'll often communicate with hardware devices through *memory mapping*, where the internal state of the device is made available in carefully chosen regions in memory. For example, while your computer starts up, the memory address range 0xA0000–0xBFFFF maps to a crude graphics rendering pipeline; writes to individual bytes in that range will change particular pixels (or blocks, depending on the mode) on the screen.

When you're interacting with device-mapped memory, the device may implement custom behavior for each memory access to that region of memory, so the assumptions your CPU and compiler make about regular memory loads and stores may no longer hold. For instance, it is common for hardware devices to have memory-mapped registers that are modified when they're read, meaning the reads have side effects. In such cases, the compiler can't safely elide a memory store operation if you read the same memory address twice in a row!

A similar issue arises when program execution is suddenly diverted in ways that aren't represented in the code and thus that the compiler cannot expect. Execution might be diverted if there is no underlying operating system to handle processor exceptions or interrupts, or if a process receives a signal that interrupts execution. In those cases, the execution of the active segment of code is stopped, and the CPU starts executing instructions in the event handler for whatever event triggered the diversion instead. Normally, since the compiler can anticipate all possible executions, it arranges its optimizations so that executions cannot observe when operations have been performed out of order or optimized away. However, since the compiler can't predict these exceptional jumps, it also cannot plan for them to be oblivious to its optimizations, so these event handlers might actually observe instructions that have run in a different order than those in the original program code.

To deal with these exceptional situations, Rust provides *volatile* memory operations that cannot be elided or reordered with respect to other volatile operations. These operations take the form of std::ptr::read_volatile and std::ptr::write_volatile. Volatile operations are exactly the right fit for accessing memory-mapped hardware resources: they map directly to memory access operations with no compiler trickery, and the guarantee that volatile operations aren't reordered relative to one another ensures that hardware operations with possible side effects don't happen out of order even when they would normally look interchangeable (such as a load of one address and a store to a different address). The no-reordering guarantee also helps the exceptional execution situation, as long as any code that touches memory accessed in an exceptional context uses only volatile memory operations.

> **NOTE** *There is also a std::sync::atomic::compiler_fence function that prevents the compiler from reordering non-volatile memory accesses. You'll very rarely need a compiler fence, but its documentation is an interesting read.*

INCLUDING ASSEMBLY CODE

These days, you rarely need to drop down to writing assembly code to accomplish any given task. But for low-level hardware programming where you need to initialize CPUs at boot or issue strange instructions to manipulate memory mappings, assembly code is still sometimes required. At the time of writing, there is an RFC and a mostly complete implementation of inline assembly syntax on nightly Rust, but nothing has been stabilized yet, so I won't discuss the syntax in this book.

It's still possible to write assembly on stable Rust—you just need to get a little creative. Specifically, remember build scripts from Chapter 11? Well, Cargo build scripts can emit certain special directives to standard output to augment Cargo's standard build process, including `cargo:rustc-link-lib=static=` *xyz* to link the static library file *libxyz.a* into the final binary, and `cargo:rustc-link-search:` */some/path* to add */some/path* to the search path for link objects. Using those, we can add a *build.rs* to the project that compiles a standalone assembly file (*.s*) to an object file (*.o*) using the target platform's compiler and then repackages it into a static archive (*.a*) using the appropriate archiving tool (usually ar). The project then emits those two Cargo directives, pointing at where it placed the static archive—probably in `OUT_DIR`—and we're off to the races! If the target platform doesn't change, you can even include the precompiled *.a* when publishing your crate so that consumers don't need to rebuild it.

Misuse-Resistant Hardware Abstraction

Rust's type system excels at encapsulating unsafe, hairy, and otherwise unpleasant code behind safe, ergonomic interfaces. Nowhere is that more important than in the infamously complex world of low-level systems programming, littered with magic hardware-defined values pulled from obscure manuals and mysterious undocumented assembly instruction incantations to get devices into just the right state. And all that in a space where a runtime error might crash more than just a user program!

In `no_std` programs, it is immensely important to use the type system to make illegal states impossible to represent, as we discussed in Chapter 3. If certain combinations of register values cannot occur at the same time, then create a single type whose type parameters indicate the current state of the relevant registers, and implement only legal transitions on it, like we did for the rocket example in Listing 3-2.

NOTE *Make sure to also review the advice from Chapter 3 on API design—all of that applies in the context of no_std programs as well!*

For example, consider a pair of registers where at most one register should be "on" at any given point in time. Listing 12-2 shows how you can represent that in a (single-threaded) program in a way makes it impossible to write code that violates that invariant.

```
// raw register address -- private submodule
mod registers;
pub struct On;
pub struct Off;
pub struct Pair<R1, R2>(PhantomData<(R1, R2)>);
impl Pair<Off, Off> {
    pub fn get() -> Option<Self> {
```

```
                    static mut PAIR_TAKEN: bool = false;
                    if unsafe { PAIR_TAKEN } {
                        None
                    } else {
                        // Ensure initial state is correct.
                        registers::off("r1");
                        registers::off("r2");
                        unsafe { PAIR_TAKEN = true };
                        Some(Pair(PhantomData))
                    }
                }

                pub fn first_on(self) -> Pair<On, Off> {
                    registers::set_on("r1");
                    Pair(PhantomData)
                }
                // .. and inverse for -> Pair<Off, On>
            }
            impl Pair<On, Off> {
                pub fn off(self) -> Pair<Off, Off> {
                    registers::set_off("r1");
                    Pair(PhantomData)
                }
            }
            // .. and inverse for Pair<Off, On>
```

Listing 12-2: Statically ensuring correct operation

There are a few noteworthy patterns in this code. The first is that we ensure only a single instance of `Pair` ever exists by checking a private static Boolean in its only constructor and making all methods consume `self`. We then ensure that the initial state is valid and that only valid state transitions are possible to express, and therefore the invariant must hold globally.

The second noteworthy pattern in Listing 12-2 is that we use `PhantomData` to take advantage of zero-sized types and represent runtime information statically. That is, at any given point in the code the types tell us what the runtime state *must* be, and therefore we don't need to track or check any state related to the registers at runtime. There's no need to check that r2 isn't already on when we're asked to enable r1, since the types prevent writing a program in which that is the case.

Cross-Compilation

Usually, you'll write no_std programs on a computer with a full-fledged operating system running and all the niceties of modern hardware, but ultimately run it on a dinky hardware device with 9¾ bits of RAM and a sock for a CPU. That calls for *cross-compilation*—you need to compile the code in your development environment, but compile it *for* the sock. That's not the only context in which cross-compilation is important, though. For example, it's increasingly common to have one build pipeline produce binary artifacts for all consumer platforms rather than trying to have a build pipeline for every platform your consumers may be using, and that means using cross-compilation.

Cross-compiling involves two platforms: the *host* platform and the *target* platform. The host platform is the one doing the compiling, and the target platform is the one that will eventually run the output of the compilation. We specify platforms as *target triples*, which take the form *machine-vendor-os*. The *machine* part dictates the machine architecture the code will run on, such as x86_64, armv7, or wasm32, and tells the compiler what instruction set to use for the emitted machine code. The *vendor* part generally takes the value of pc on Windows, apple on macOS and iOS, and unknown everywhere else, and doesn't affect compilation in any meaningful way; it's mostly irrelevant and can even be left out. The *os* part tells the compiler what format to use for the final binary artifacts, so a value of linux dictates Linux *.so* files, windows dictates Windows *.dll* files, and so on.

To tell Cargo to cross-compile, you simply pass it the --target <target triple> argument with your triple of choice. Cargo will then take care of forwarding that information to the Rust compiler so that it generates binary artifacts that will work on the given target platform. Cargo will also take care to use the appropriate version of the standard library for that platform—after all, the standard library contains a lot of conditional compilation directives (using #[cfg(...)]) so that the right system calls get invoked and the right architecture-specific implementations are used, so we can't use the standard library for the host platform on the target.

The target platform also dictates what components of the standard library are available. For example, while x86_64-unknown-linux-gnu includes the full std library, something like thumbv7m-none-eabi does no, and doesn't even define an allocator, so if you use alloc without defining one explicitly, you'll get a build error. This comes in handy for testing that code you write *actually* doesn't require std (recall that even with #![no_std] you can still have use std::, since no_std opts out of only the std prelude). If you have your continuous integration pipeline build your crate with --target thumbv7m-none-eabi, any attempt to access components from anything but core will trigger a build failure. Crucially, this will also check that your crate doesn't accidentally bring in dependencies that themselves use items from std (or alloc).

PLATFORM SUPPORT

The standard Rust installer, Rustup, doesn't install the standard library for all the target triples that Rust supports by default. That would be a waste of space and bandwidth. Instead, you have to use the command `rustup target add` to install the appropriate standard library versions for additional targets. If no version of the standard library exists for your target platform, you'll have to compile it from source yourself by adding the `rust-src` Rustup component and using Cargo's (currently unstable) `build-std` feature to also build `std` (and/or `core` and `alloc`) when building any crate.

If your target is not supported by the Rust compiler—that is, if `rustc` doesn't even know about your target triple—you'll have to go one step further and teach `rustc` about the properties of the triple using a custom target specification. How you do that is both currently unstable and beyond the scope of this book, but a search for "custom target specification json" is a good place to start.

Summary

In this chapter, we've covered what lies beneath the standard library—or, more precisely, beneath std. We've gone over what you get with core, how you can extend your non-std reach with alloc, and what the (tiny) Rust runtime adds to your programs to make fn main work. We've also taken a look at how you can interact with device-mapped memory and otherwise handle the unorthodox execution patterns that can happen at the very lowest level of hardware programming, and how to safely encapsulate at least some of the oddities of hardware in the Rust type system. Next, we'll move from the very small to the very large by discussing how to navigate, understand, and maybe even contribute to the larger Rust ecosystem.

13

THE RUST ECOSYSTEM

Programming rarely happens in a vacuum these days—nearly every Rust crate you build is likely to take dependencies on *some* code that wasn't written by you. Whether this trend is good, bad, or a little of both is a subject of heavy debate, but either way, it's a reality of today's developer experience.

In this brave new interdependent world, it's more important than ever to have a solid grasp of what libraries and tools are available and to stay up to date on the latest and greatest of what the Rust community has to offer. This chapter is dedicated to how you can leverage, track, understand, and contribute back to the Rust ecosystem. Since this is the final chapter, in the closing section I'll also provide some suggestions of additional resources you can explore to continue developing your Rust skills.

What's Out There?

Despite its relative youth, Rust already has an ecosystem large enough that it's hard to keep track of everything that's available. If you know what you want, you may be able to search your way to a set of appropriate crates and then use download statistics and superficial vibe-checks on each crate's repository to determine which may make for reasonable dependencies. However, there's also a plethora of tools, crates, and general language features that you might not necessarily know to look for that could potentially save you countless hours and difficult design decisions.

In this section, I'll go through some of the tools, libraries, and Rust features I have found helpful over the years in the hopes that they may come in useful for you at some point too!

Tools

First off, here are some Rust tools I find myself using regularly that you should add to your toolbelt:

cargo-deny

Provides a way to lint your dependency graph. At the time of writing, you can use `cargo-deny` to allow only certain licenses, deny-list crates or specific crate versions, detect dependencies with known vulnerabilities or that use Git sources, and detect crates that appear multiple times with different versions in the dependency graph. By the time you're reading this, there may be even more handy lints in place.

cargo-expand

Expands macros in a given crate and lets you inspect the output, which makes it much easier to spot mistakes deep down in macro transcribers or procedural macros. `cargo-expand` is an invaluable tool when you're writing your own macros.

cargo-hack

Helps you check that your crate works with any combination of features enabled. The tool presents an interface similar to that of Cargo itself (like `cargo check`, `build`, and `test`) but gives you the ability to run a given command with all possible combinations (the *powerset*) of the crate's features.

cargo-llvm-lines

Analyzes the mapping from Rust code to the intermediate representation (IR) that's passed to the part of the Rust compiler that actually generates machine code (LLVM), and tells you which bits of Rust code produce the largest IR. This is useful because a larger IR means longer compile times, so identifying what Rust code generates a bigger IR (due to, for example, monomorphization) can highlight opportunities for reducing compile times.

cargo-outdated

Checks whether any of your dependencies, either direct or transitive, have newer versions available. Crucially, unlike `cargo update`, it even tells you about new major versions, so it's an essential tool for checking if you're missing out on newer versions due to an outdated major version specifier. Just keep in mind that bumping the major version of a dependency may be a breaking change for your crate if you expose that dependency's types in your interface!

cargo-udeps

Identifies any dependencies listed in your *Cargo.toml* that are never actually used. Maybe you used them in the past but they've since become redundant, or maybe they should be moved to `dev-dependencies`; whatever the case, this tool helps you trim down bloat in your dependency closure.

While they're not specifically tools for developing Rust, I highly recommend `fd` and `ripgrep` too—they're excellent improvements over their predecessors `find` and `grep` and also happen to be written in Rust themselves. I use both every day.

Libraries

Next up are some useful but lesser-known crates that I reach for regularly, and that I suspect I will continue to depend on for a long time:

bytes

Provides an efficient mechanism for passing around subslices of a single piece of contiguous memory without having to copy or deal with lifetimes. This is great in low-level networking code where you may need multiple views into a single chunk of bytes, and copying is a no-no.

criterion

A statistics-driven benchmarking library that uses math to eliminate noise from benchmark measurements and reliably detect changes in performance over time. You should almost certainly be using it if you're including micro-benchmarks in your crate.

cxx

Provides a safe and ergonomic mechanism for calling C++ code from Rust and Rust code from C++. If you're willing to invest some time into declaring your interfaces more thoroughly in advance in exchange for much nicer cross-language compatibility, this library is well worth your attention.

flume

Implements a multi-producer, multi-consumer channel that is faster, more flexible, and simpler than the one included with the Rust standard library. It also supports both asynchronous and synchronous operation and so is a great bridge between those two worlds.

hdrhistogram

A Rust port of the High Dynamic Range (HDR) histogram data structure, which provides a compact representation of histograms across a wide range of values. Anywhere you currently track averages or min/ max values, you should most likely be using an HDR histogram instead; it can give you much better insight into the distribution of your metrics.

heapless

Supplies data structures that do not use the heap. Instead, `heapless`'s data structures are all backed by static memory, which makes them perfect for embedded contexts or other situations in which allocation is undesirable.

itertools

Extends the `Iterator` trait from the standard library with lots of new convenient methods for deduplication, grouping, and computing powersets. These extension methods can significantly reduce boilerplate in code, such as where you manually implement some common algorithm over a sequence of values, like finding the min and max at the same time (`Itertools::minmax`), or where you use a common pattern like checking that an iterator has exactly one item (`Itertools::exactly_one`).

nix

Provides idiomatic bindings to system calls on Unix-like systems, which allows for a much better experience than trying to cobble together the C-compatible FFI types yourself when working with something like `libc` directly.

pin-project

Provides macros that enforce the pinning safety invariants for annotated types, which in turn provide a safe pinning interface to those types. This allows you to avoid most of the hassle of getting `Pin` and `Unpin` right for your own types. There's also `pin-project-lite`, which avoids the (currently) somewhat heavy dependency on the procedural macro machinery at the cost of slightly worse ergonomics.

ring

Takes the good parts from the cryptography library BoringSSL, written in C, and brings them to Rust through a fast, simple, and hard-to-misuse interface. It's a great starting point if you need to use cryptography in your crate. You've already most likely come across this in the `rustls` library, which uses `ring` to provide a modern, secure-by-default TLS stack.

slab

Implements an efficient data structure to use in place of `HashMap<Token, T>`, where `Token` is an opaque type used only to differentiate between entries in the map. This kind of pattern comes up a lot when managing resources, where the set of current resources must be managed centrally but individual resources must also be accessible somehow.

static_assertions

Provides static assertions—that is, assertions that are evaluated at, and thus may fail at, compile time. You can use it to assert things like that a type implements a given trait (like `Send`) or is of a given size. I highly recommend adding these kinds of assertions for code where those guarantees are likely to be important.

structopt

Wraps the well-known argument parsing library `clap` and provides a way to describe your application's command line interface entirely using the Rust type system (plus macro annotations). When you parse your application's arguments, you get a value of the type you defined, and you thus get all the type checking benefits, like exhaustive matching and IDE auto-complete.

thiserror

Makes writing custom enumerated error types, like the ones we discussed in Chapter 4, a joy. It takes care of implementing the recommended traits and following the established conventions and leaves you to define just the critical bits that are unique to your application.

tower

Effectively takes the function signature `async fn(Request) -> Response` and implements an entire ecosystem on top of it. At its core is the `Service` trait, which represents a type that can turn a request into a response (something I suspect may make its way into the standard library one day). This is a great abstraction to build anything that looks like a service on top of.

tracing

Provides all the plumbing needed to efficiently trace the execution of your applications. Crucially, it is agnostic to the types of events you're tracing and what you want to do with those events. This library can be used for logging, metrics collection, debugging, profiling, and obviously tracing, all with the same machinery and interfaces.

Rust Tooling

The Rust toolchain has a few features up its sleeve that you may not know to look for. These are usually for very specific use cases, but if they match yours, they can be lifesavers!

Rustup

Rustup, the Rust toolchain installer, does its job so efficiently that it tends to fade into the background and get forgotten about. You'll occasionally use it to update your toolchain, set a directory override, or install a component, but that's about it. However, Rustup supports one very handy trick that it's worthwhile to know about: the toolchain override shorthand. You can pass +toolchain as the first argument to any Rustup-managed binary, and the binary will work as if you'd set an override for the given toolchain, run the command, and then reset the override back to what it was previously. So, cargo +nightly miri will run Miri using the nightly toolchain, and cargo +1.53.0 check will check if the code compiles with Rust 1.53.0. The latter comes in particularly handy for checking that you haven't broken your minimum supported Rust version contract.

Rustup also has a neat subcommand, doc, that opens a local copy of the Rust standard library documentation for the current version of the Rust compiler in your browser. This is invaluable if you're developing on the go without an internet connection!

Cargo

Cargo also has some handy features that aren't always easy to discover. The first of these is cargo tree, a Cargo subcommand built right into Cargo itself for inspecting a crate's dependency graph. This command's primary purpose is to print the dependency graph as a tree. This can be useful on its own, but where cargo tree really shines is through the --invert option: it takes a crate identifier and produces an inverted tree showing all the dependency paths from the current crate that bring in that dependency. So, for example, cargo tree -i rand will print all of the ways in which the current crate depends on any version of rand, including through transitive dependencies. This is invaluable if you want to eliminate a dependency, or a particular version of a dependency, and wonder why it still keeps being pulled in. You can also pass the -e features option to include information about why each Cargo feature of the crate in question is enabled.

Speaking of Cargo subcommands, it's really easy to write your own, whether for sharing with other people or just for your own local development. When Cargo is invoked with a subcommand it doesn't recognize, it checks whether a program by the name cargo-$subcommand exists. If it does, Cargo invokes that program and passes it any arguments that were passed on the command line—so, `cargo foo bar` will invoke `cargo-foo` with the argument bar. Cargo will even integrate this command with `cargo help` by translating `cargo help foo` into a call to `cargo-foo --help`.

As you work on more Rust projects, you may notice that Cargo (and Rust more generally) isn't exactly forgiving when it comes to disk space. Each project gets its own target directory for its compilation artifacts, and over time you end up accumulating several identical copies of compiled artifacts for common dependencies. Keeping artifacts for each project separate is a sensible choice, as they aren't necessarily compatible across projects (say, if one project uses different compiler flags than another). But in most developer environments, sharing build artifacts is entirely reasonable and can save a fair amount of compilation time when switching between projects. Luckily, configuring Cargo to share build artifacts is simple: just set [build] target in your *~/.cargo/config.toml* file to the directory you want those shared artifacts to go in, and Cargo will take care of the rest. No more target directories in sight! Just make sure you clean out that directory every now and again too, and be aware that `cargo clean` will now clean *all* of your projects' build artifacts.

NOTE *Using a shared build directory can cause problems for projects that assume that compiler artifacts will always be under the* target/ *subdirectory, so watch out for that. Also note that if a project* does *use different compiler flags, you'll end up recompiling affected dependencies every time you move into or out of that project. In such cases, you're best off overriding the target directory in that project's Cargo configuration to a distinct location.*

Finally, if you ever feel like Cargo is taking a suspiciously long time to build your crate, you can reach for the currently unstable Cargo `-Ztimings` flag. Running Cargo with that flag outputs information about how long it took to process each crate, how long build scripts took to run, what crates had to wait for what other crates to finish compiling, and tons of other useful metrics. This might highlight a particularly slow dependency chain that you can then work to eliminate, or reveal a build script that compiles a native dependency from scratch that you can make use system libraries instead. If you want to dive even deeper, there's also `rustc -Ztime-passes`, which emits information about where time is spent inside of the compiler for each crate—though that information is likely only useful if you're looking to contribute to the compiler itself.

rustc

The Rust compiler also has some lesser-known features that can prove useful to enterprising developers. The first is the currently unstable `-Zprint-type-sizes` argument, which prints the sizes of all the types in the current crate. This produces a lot of information for all but the tiniest crates but is immensely valuable when trying to determine the source of unexpected time spent in calls to `memcpy` or to find ways to reduce memory use when allocating lots of objects of a particular type. The `-Zprint-type-sizes` argument

The Rust Ecosystem   **229**

also displays the computed alignment and layout for each type, which may point you to places where turning, say, a usize into a u32 could have a significant impact on a type's in-memory representation. After you debug a particular type's size, alignment, and layout, I recommend adding static assertions to make sure that they don't regress over time. You may also be interested in the variant_size_differences lint, which issues a warning if a crate contains enum types whose variants significantly differ in size.

<blockquote>
NOTE   To call rustc with particular flags, you have a few options: you can either set them in the RUSTFLAGS environment variable or [build] rustflags in your .cargo/config.toml to have them apply to every invocation of rustc from Cargo, or you can use cargo rustc, which will pass any arguments you provide only to the rustc invocation for the current crate.
</blockquote>

If your profiling samples look weird, with stack frames reordered or entirely missing, you could also try -Cforce-frame-pointers = yes. Frame pointers provide a more reliable way to unwind the stack—which is done a lot during profiling—at the cost of an extra register being used for function calls. Even though stack unwinding *should* work fine with just regular debug symbols enabled (remember to set debug = true when using the release profile), that's not always the case, and frame pointers may take care of any issues you do encounter.

The Standard Library

The Rust standard library is generally considered to be small compared to those of other programming languages, but what it lacks in breadth, it makes up for in depth; you won't find a web server implementation or an X.509 certificate parser in Rust's standard library, but you will find more than 40 different methods on the Option type alongside over 20 trait implementations. For the types it does include, Rust does its best to make available any relevant functionality that meaningfully improves ergonomics, so you avoid all that verbose boilerplate that can so easily arise otherwise. In this section, I'll present some types, macros, functions, and methods from the standard library that you may not have come across before, but that can often simplify or improve (or both) your code.

Macros and Functions

Let's start off with a few free-standing utilities. First up is the **write!** macro, which lets you use format strings to write into a file, a network socket, or anything else that implements Write. You may already be familiar with it—but one little-known feature of write! is that it works with both std::io::Write and std::fmt::Write, which means you can use it to write formatted text directly into a String. That is, you can write use std::fmt::Write; write!(&mut s, "{}+1={}", x, x + 1); to append the formatted text to the String s!

The **iter::once** function takes any value and produces an iterator that yields that value once. This comes in handy when calling functions that take iterators if you don't want to allocate, or when combined with `Iterator::chain` to append a single item to an existing iterator.

We briefly talked about `mem::replace` in Chapter 1, but it's worth bringing it up again in case you missed it. This function takes an exclusive reference to a T and an owned T, swaps the two so that the referent is now the owned T, and returns ownership of the previous referent. This is useful when you need to take ownership of a value in a situation where you have only an exclusive reference, such as in implementations of `Drop`. See also `mem::take` for when `T: Default`.

Types

Next, let's look at some handy standard library types. The **BufReader** and **BufWriter** types are a must for I/O operations that issue many small read or write calls to the underlying I/O resource. These types wrap the respective underlying `Read` or `Write` and implement `Read` and `Write` themselves, but they additionally buffer the operations to the I/O resource such that many small reads do only one large read, and many small writes do only one large write. This can significantly improve performance as you don't have to cross the system call barrier into the operating system as often.

The `Cow` type, mentioned in Chapter 3, is useful when you want flexibility in what types you hold or need flexibility in what you return. You'll rarely use `Cow` as a function argument (recall that you should let the caller allocate if necessary), but it's invaluable as a return type as it allows you to accurately represent the return types of functions that may or may not allocate. It's also a perfect fit for types that can be used as inputs *or* outputs, such as core types in RPC-like APIs. Say we have a type `EntityIdentifier` like in Listing 13-1 that is used in an RPC service interface.

```
struct EntityIdentifier {
    namespace: String,
    name: String,
}
```

Listing 13-1: A representation of a combined input/output type that requires allocation

Now imagine two methods: get_entity takes an `EntityIdentifier` as an argument, and find_by returns an `EntityIdentifier` based on some search parameters. The get_entity method requires only a reference since the identifier will (presumably) be serialized before being sent to the server. But for find_by, the entity will be deserialized from the server response and must therefore be represented as an owned value. If we make get_entity take &EntityIdentifier, it will mean callers must still allocate owned Strings to call get_entity even though that's not required by the interface, since it's required to construct an `EntityIdentifier` in the first place! We could instead introduce a separate type for get_entity, EntityIdenifierRef, that holds only &str types, but then we'd have two types to represent one thing. Cow to the rescue! Listing 13-2 shows an `EntityIdentifier` that instead holds Cows internally.

```
struct EntityIdentifier<'a> {
    namespace: Cow<'a, str>,
    name: Cow<'a str>,
}
```

Listing 13-2: A representation of a combined input/output type that does not require allocation

With this construction, get_entity can take any EntityIdentifier<'_>, which allows the caller to use just references to call the method. And find_by can return EntityIdentifier<'static>, where all the fields are Cow::Owned. One type shared across both interfaces, with no unnecessary allocation requirements!

NOTE *If you implement a type this way, I recommend you also provide an into_owned method that turns an <'a> instance into a <'static> instance by calling Cow::into_owned on all the fields. Otherwise, users will have no way to make longer-lasting clones of your type when all they have is an <'a>.*

The **std::sync::Once** type is a synchronization primitive that lets you run a given piece of code exactly once, at initialization time. This is great for initialization that's part of an FFI where the library on the other side of the FFI boundary requires that the initialization is performed only once.

The **VecDeque** type is an oft-neglected member of std::collections that I find myself reaching for surprisingly often—basically, whenever I need a stack or a queue. Its interface is similar to that of a Vec, and like Vec its in-memory representation is a single chunk of memory. The difference is that VecDeque keeps track of both the start and end of the actual data in that single allocation. This allows constant-time push and pop from *either* side of the VecDeque, meaning it can be used as a stack, as a queue, or even both at the same time. The cost you pay is that the values are no longer necessarily contiguous in memory (they may have wrapped around), which means that VecDeque<T> does not implement AsRef<[T]>.

Methods

Let's round off with a rapid-fire look at some neat methods. First up is **Arc::make_mut**, which takes a &mut Arc<T> and gives you a &mut T. If the Arc is the last one in existence, it gives you the T that was behind the Arc; otherwise, it allocates a new Arc<T> that holds a clone of the T, swaps that in for the currently referenced Arc, and then gives &mut to the T in the new singleton Arc.

The **Clone::clone_from** method is an alternative form of .clone() that lets you reuse an instance of the type you clone rather than allocate a new one. In other words, if you already have an x: T, you can do x.clone_from(y) rather than x = y.clone(), and you might save yourself some allocations.

std::fmt::Formatter::debug_* is by far the easiest way to implement Debug yourself if #[derive(Debug)] won't work for your use case, such as if you want to include only some fields or expose information that isn't exposed by the

Debug implementations of your type's fields. When implementing the fmt method of Debug, simply call the appropriate debug_ method on the Formatter that's passed in (debug_struct or debug_map, for example), call the included methods on the resulting type to fill in details about the type (like field to add a field or entries to add a key/value entry), and then call finish.

Instant::elapsed returns the Duration since an Instant was created. This is much more concise than the common approach of creating a new Instant and subtracting the earlier instance.

Option::as_deref takes an Option<P> where P: Deref and returns Option<&P::Target> (there's also an as_deref_mut method). This simple operation can make functional transformation chains that operate on Option much cleaner by avoiding the inscrutable .as_ref().map(|r| &**r).

Ord::clamp lets you take any type that implements Ord and clamp it between two other values of a given range. That is, given a lower limit min and an upper limit max, x.clamp(min, max) returns min if x is less than min, max if x is greater than max, and x otherwise.

Result::transpose and its counterpart Option::transpose invert types that nest Result and Option. That is, transposing a Result<Option<T>, E> gives an Option<Result<T, E>>, and vice versa. When combined with ?, this operation can make for cleaner code when working with Iterator::next and similar methods in fallible contexts.

Vec::swap_remove is Vec::remove's faster twin. Vec::remove preserves the order of the vector, which means that to remove an element in the middle, it must shift all the later elements in the vector down by one. This can be very slow for large vectors. Vec::swap_remove, on the other hand, swaps the to-be-removed element with the last element and then truncates the vector's length by one, which is a constant-time operation. Be aware, though, that it will shuffle your vector around and thus invalidate old indexes!

Patterns in the Wild

As you start exploring codebases that aren't your own, you'll likely come across a couple of common Rust patterns that we haven't discussed in the book so far. Knowing about them will make it easier to recognize them, and thus understand their purpose, when you do encounter them. You may even find use for them in your own codebase one day!

Index Pointers

Index pointers allow you to store multiple references to data within a data structure without running afoul of the borrow checker. For example, if you want to store a collection of data so that it can be efficiently accessed in more than one way, such as by keeping one HashMap keyed by one field and one keyed by a different field, you don't want to store the underlying data multiple times too. You could use Arc or Rc, but they use dynamic reference counting that introduces unnecessary overhead, and the extra bookkeeping requires you to store additional bytes per entry. You could use references, but the lifetimes become difficult if not impossible to manage because the data and the references live in the same data structure (it's a self-referential data structure, as we discussed in Chapter 8). You could use raw pointers combined with Pin to ensure the pointers remain valid, but that introduces a lot of complexity as well as unsafety you then need to carefully consider.

Most crates use index pointers—or, as I like to call them, *indeferences*—instead. The idea is simple: store each data entry in some indexable data structure like a Vec, and then store just the index in a derived data structure. To then perform an operation, first use the derived data structure to efficiently find the data index, and then use the index to retrieve the referenced data. No lifetimes needed—and you can even have cycles in the derived data representation if you wish!

The indexmap crate, which provides a HashMap implementation where the iteration order matches the map insertion order, provides a good example of this pattern. The implementation has to store the keys in two places, both in the map of keys to values and in the list of all the keys, but it obviously doesn't want to keep two copies in case the key type itself is large. So, it uses index pointers. Specifically, it keeps all the key/value pairs in a single Vec and then keeps a mapping from key hashes to Vec indexes. To iterate over all the elements of the map, it just walks the Vec. To look up a given key, it hashes that key, looks that hash up in the mapping, which yields the key's index in the Vec (the index pointer), and then uses that to get the key's value from the Vec.

The petgraph crate, which implements graph data structures and algorithms, also uses this pattern. The crate stores one Vec of all node values and another of all edge values and then only ever uses the indexes into those Vecs to refer to a node or edge. So, for example, the two nodes associated with an edge are stored in that edge simply as two u32s, rather than as references or reference-counted values.

The trick lies in how you support deletions. To delete a data entry, you first need to search for its index in all of the derived data structures and remove the corresponding entries, and then you need to remove the data from the root data store. If the root data store is a Vec, removing the entry will also change the index of one other data entry (when using swap_remove), so you then need to go update all the derived data structures to reflect the new index for the entry that moved.

Drop Guards

Drop guards provide a simple but reliable way to ensure that a bit of code runs even in the presence of panics, which is often essential in unsafe code. An example is a function that takes a closure f: FnOnce and executes it under mutual exclusion using atomics. Say the function uses compare_exchange (discussed in Chapter 10) to set a Boolean from false to true, calls f, and then sets the Boolean back to false to end the mutual exclusion. But consider what happens if f panics—the function will never get to run its cleanup, and no other call will be able to enter the mutual exclusion section ever again.

It's possible to work around this using catch_unwind, but drop guards provide an alternative that is often more ergonomic. Listing 13-3 shows how, in our current example, we can use a drop guard to ensure the Boolean always gets reset.

```
fn mutex(lock: &AtomicBool, f: impl FnOnce()) {
    // .. while lock.compare_exchange(false, true).is_err() ..
    struct DropGuard<'a>(&'a AtomicBool);
    impl Drop for DropGuard<'_> {
        fn drop(&mut self) {
            lock.store(true, Ordering::Release);
        }
    }
    let _guard = DropGuard(lock);
    f();
}
```

Listing 13-3: Using a drop guard to ensure code gets run after an unwinding panic

We introduce the local type DropGuard that implements Drop and place the cleanup code in its implementation of Drop::drop. Any necessary state can be passed in through the fields of DropGuard. Then, we construct an instance of the guard type just before we call the function that might panic, which is f here. When f returns, whether due to a panic or because it returns normally, the guard is dropped, its destructor runs, the lock is released, and all is well.

It's important that the guard is assigned to a variable that is dropped at the end of the scope, after the user-provided code has been executed. This means that even though we never refer to the guard's variable again, it needs to be given a name, as let _ = DropGuard(lock) would drop the guard immediately—before the user-provided code even runs!

NOTE *Like catch_unwind, drop guards work only when panics unwind. If the code is compiled with panic=abort, no code gets to run after the panic.*

This pattern is frequently used in conjunction with thread locals, when library code may wish to set the thread local state so that it's valid only for the duration of the execution of the closure, and thus needs to be cleared afterwards. For example, at the time of writing, Tokio uses this pattern to provide information about the executor calling Future::poll to leaf resources like TcpStream without having to propagate that information through function signatures that are visible to users. It'd be no good if the thread local state continued to indicate that a particular executor thread was active even after Future::poll returned due to a panic, so Tokio uses a drop guard to ensure that the thread local state is reset.

NOTE *You'll often see Cell or Rc<RefCell> used in thread locals. This is because thread locals are accessible only through shared references, since a thread might access a thread local again that it is already referencing somewhere higher up in the call stack. Both types provide interior mutability without incurring much overhead because they're intended only for single-threaded use, and so are ideal for this use case.*

Extension Traits

Extension traits allow crates to provide additional functionality to types that implement a trait from a different crate. For example, the itertools crate provides an extension trait for Iterator, which adds a number of convenient shortcuts for common (and not so common) iterator operations. As another example, tower provides ServiceExt, which adds several more ergonomic operations to wrap the low-level interface in the Service trait from tower-service.

Extension traits tend to be useful either when you do not control the base trait, as with Iterator, or when the base trait lives in a crate of its own so that it rarely sees breaking releases and thus doesn't cause unnecessary ecosystem splits, as with Service.

An extension trait extends the base trait it is an extension of (trait ServiceExt: Service) and consists solely of provided methods. It also comes with a blanket implementation for any T that implements the base trait (impl<T> ServiceExt for T where T: Service {}). Together, these conditions ensure that the extension trait's methods are available on anything that implements the base trait.

Crate Preludes

In Chapter 12, we talked about the standard library prelude that makes a number of types and traits automatically available without you having to write any use statements. Along similar lines, crates that export multiple types, traits, or functions that you'll often use together sometimes define their own prelude in the form of a module called prelude, which re-exports some particularly common subset of those types, traits, and functions. There's nothing magical about that module name, and it doesn't get used automatically, but it serves as a signal to users that they likely want to add use *somecrate*::prelude::* to files that want to use the crate in question. The * is a *glob import* and tells Rust to use all publicly available items from the indicated module. This can save quite a bit of typing when the crate has a lot of items you'll usually need to name.

NOTE *Items used through * have lower precedence than items that are used explicitly by name. This is what allows you to define items in your own crate that overlap with what's in the standard library prelude without having to specify which one to use.*

Preludes are also great for crates that expose a lot of extension traits, since trait methods can be called only if the trait that defines them is in scope. For example, the diesel crate, which provides ergonomic access to relational databases, makes extensive use of extension traits so you can write code like:

```
posts.filter(published.eq(true)).limit(5).load::<Post>(&connection)
```

This line will work only if all the right traits are in scope, which the prelude takes care of.

In general, you should be careful when adding glob imports to your code, as they can potentially turn additions to the indicated module into backward-incompatible changes. For example, if someone adds a new trait to a module you glob-import from, and that new trait makes a method foo available on a type that already had some other foo method, code that calls foo on that type will no longer compile as the call to foo is now ambiguous. Interestingly enough, while the existence of glob imports makes any module addition a technically breaking change, the Rust RFC on API evolution (RFC 1105; see *https://rust-lang.github.io/rfcs/1105-api-evolution.html*) does *not* require a library to issue a new major version for such a change. The RFC goes into great detail about why, and I recommend you read it, but the gist is that minor releases are allowed to require minimally invasive changes to dependents, like having to add type annotations in edge cases, because otherwise a large fraction of changes would require new major versions despite being very unlikely to actually break any consumers.

Specifically in the case of preludes, using glob imports is usually fine when recommended by the vending crate, since its maintainers know that their users will use glob imports for the prelude module and thus will take that into account when deciding whether a change requires a major version bump.

Staying Up to Date

Rust, being such a young language, is evolving rapidly. The language itself, the standard library, the tooling, and the broader ecosystem are all still in their infancy, and new developments happen every day. While staying on top of all the changes would be infeasible, it's worth your time to keep up with significant developments so that you can take advantage of the latest and greatest features in your projects.

For monitoring improvements to Rust itself, including new language features, standard library additions, and core tooling upgrades, the official Rust blog at *https://blog.rust-lang.org/* is a good, low-volume place to start. It mainly features announcements for each new Rust release. I recommend you make a habit of reading these, as they tend to include interesting tidbits that will slowly but surely deepen your knowledge of the language. To dig a little deeper, I highly recommend reading the detailed changelogs for Rust and Cargo as well (links can usually be found near the bottom of each release announcement). The changelogs surface changes that weren't large enough to warrant a paragraph in the release notes but that may be just what you need two weeks from now. For a less frequently updated news source, check in on *The Edition Guide* at *https://doc.rust-lang.org/edition-guide/*, which outlines what's new in each Rust edition. Rust editions tend to be released every three years.

NOTE *Clippy is often able to tell you when you can take advantage of a new language or standard library feature—always enable Clippy!*

If you're curious about how Rust itself is developed, you may also want to subscribe to the *Inside Rust* blog at *https://blog.rust-lang.org/inside-rust/*. It includes updates from the various Rust teams, as well as incident reports, larger change proposals, edition planning information, and the like. To get involved in Rust development yourself—which I highly encourage, as it's a lot of fun and a great learning experience—you can check out the various Rust working groups at *https://www.rust-lang.org/governance/*, which each focus on improving a specific aspect of Rust. Find one that appeals to you, check in with the group wherever it meets and ask how you may be able to help. You can also join the community discussion about Rust internals over at *https://internals.rust-lang.org/*; this is another great way to get insight into the thought that goes into every part of Rust's design and development.

As is the case for most programming languages, much of Rust's value is derived from its community. Not only do the members of the Rust community constantly develop new work-saving crates and discover new Rust-specific techniques and design patterns, but they also collectively and continuously help one another understand, document, and explain how to take best advantage of the Rust language. Everything I have covered in this book, and much more, has already been discussed by the community in thousands of comment threads, blog posts, and Twitter and Discord conversations. Dipping into these discussions even just once in a while is almost guaranteed to show you new things about a language feature, a technique, or a crate that you didn't already know.

The Rust community lives in a lot of places, but some good places to start are the Users forum (*https://users.rust-lang.org/*), the Rust subreddit (*https://www.reddit.com/r/rust/*), the Rust Community Discord (*https://discord.gg/rust-lang-community*), and the Rust Twitter account (*https://twitter.com/rustlang*). You don't have to engage with all of these, or all of the time—pick one you like the vibe of, and check in occasionally!

A great single location for staying up to date with ongoing developments is the *This Week in Rust* blog (*https://this-week-in-rust.org/*), a "weekly summary of [Rust's] progress and community." It links to official announcements and changelogs as well as popular community discussions and resources, interesting new crates, opportunities for contributions, upcoming Rust events, and Rust job opportunities. It even lists interesting language RFCs and compiler PRs, so this site truly has it all! Discerning what information is valuable to you and what isn't may be a little daunting, but even just scrolling through and clicking occasional links that appear interesting is a good way to keep a steady stream of new Rust knowledge trickling into your brain.

NOTE *Want to look up when a particular feature landed on stable? Can I Use... (https://caniuse.rs/) has you covered.*

What Next?

So, you've read this book front to back, absorbed all the knowledge it imparts, and are still hungry for more? Great! There are a number of other excellent resources out there for broadening and deepening your knowledge and understanding of Rust, and in this very final section I'll give you a survey of some of my favorites so that you can keep learning. I've divided them into subsections based on how different people prefer to learn so that you can find resources that'll work for you.

A challenge with learning on your own, especially in the beginning, is that progress is hard to perceive. Implementing even the simplest of things can take an outsized amount of time when you have to constantly refer to documentation and other resources, ask for help, or debug to learn how some aspect of Rust works. All of that non-coding work can make it seem like you're treading water and not really improving. But you're learning, *which is progress in and of itself—it's just harder to notice and appreciate.*

Learn by Watching

Watching experienced developers code is essentially a life hack to remedy the slow starting phase of solo learning. It allows you to observe the process of designing and building while utilizing someone else's experience. Listening to experienced developers articulate their thinking and explain tricky concepts or techniques as they come up can be an excellent alternative to struggling through problems on your own. You'll also pick up a variety of auxiliary knowledge like debugging techniques, design patterns, and best practices. Eventually you will have to sit down and do things yourself—it's the only way to check that you actually understand what you've observed—but piggybacking on the experience of others will almost certainly make the early stages more pleasant. And if the experience is interactive, that's even better!

So, with that said, here are some Rust video channels that I recommend:

Perhaps unsurprisingly, my own channel: *https://www.youtube.com/c/ JonGjengset/.* I have a mix of long-form coding videos and short(er) code-based theory/concept explanation videos, as well as occasional videos that dive into interesting Rust coding stories.

The *Awesome Rust Streaming* listing: *https://github.com/jamesmunns/ awesome-rust-streaming/.* This resource lists a wide variety of developers who stream Rust coding or other Rust content.

The channel of Tim McNamara, the author of *Rust in Action*: *https:// www.youtube.com/c/timClicks/.* Tim's channel, like mine, splits its time between implementation and theory, though Tim has a particular knack for creative visual projects, which makes for fun viewing.

Jonathan Turner's *Systems with JT* channel: *https://www.youtube.com/c/ SystemswithJT/.* Jonathan's videos document their work on Nushell, their take on a "new type of shell," providing a great sense of what it's like to work on a nontrivial existing codebase.

Ryan Levick's channel: *https://www.youtube.com/c/RyanLevicksVideos/.* Ryan mainly posts videos that tackle particular Rust concepts and walks through them using concrete code examples, but he also occasionally does implementation videos (like FFI for Microsoft Flight Simulator!) and deep dives into how well-known crates work under the hood.

Given that I make Rust videos, it should come as no surprise that I am a fan of this approach to teaching. But this kind of receptive or interactive learning doesn't have to come in the form of videos. Another great avenue for learning from experienced developers is pair programming. If you have a colleague or friend with expertise in a particular aspect of Rust you'd like to learn, ask if you can do a pair-programming session with them to solve a problem together!

Learn by Doing

Since your ultimate goal is to get better at writing Rust, there's no substitute for programming experience. No matter what or how many resources you learn from, you need to put that learning into practice. However, finding a good place to start can be tricky, so here I'll give some suggestions.

Before I dive into the list, I want to provide some general guidance on how to pick projects. First, choose a project that *you* care about, without worrying too much whether others care about it. While there are plenty of popular and established Rust projects out there that would love to have you as a contributor, and it's fun to be able to say "I contributed to the well-known library X," your first priority must be your own interest. Without concrete motivation, you'll quickly lose steam and find contributing to be a chore. The very best targets are projects that you use yourself and have experienced problems with—go fix them! Nothing is more satisfying than getting rid of a long-standing personal nuisance while also contributing back to the community.

Okay, so back to project suggestions. First and foremost, consider contributing to the Rust compiler and its associated tools. It's a high-quality codebase with good documentation and an endless supply of issues (you probably know of some yourself), and there are several great mentors who can provide outlines for how to approach solving issues. If you look through the issue tracker for issues marked E-easy or E-mentor, you'll likely find a good candidate quickly. As you gain more experience, you can keep leveling up to contribute to trickier parts.

If that's not your cup of tea, I recommend finding something you use frequently that's written in another language and porting it to Rust—not necessarily with the intention of replacing the original library or tool, but just because the experience will allow you to focus on writing Rust without having to spend too much time coming up with all the functionality yourself. If it turns out well, the fact that it already exists suggests that someone else also needed it, so there may be a wider audience for your port too! Data structures and command-line tools often make for great porting subjects, but find a niche that appeals to you.

Should you be more of a "build it from scratch" kind of person, I recommend looking back at your own development experience so far and thinking about similar code you've ended up writing in multiple projects (whether in Rust or in other languages). Such repetition tends to be a good signal that something is reusable and could be turned into a library. If nothing comes to mind, David Tolnay maintains a list of smaller utility crates that other Rust developers have requested at *https://github.com/dtolnay/request-for-implementation/* that may provide a source of inspiration. If you're looking for something more substantial and ambitious, there's also the Not Yet Awesome list at *https://github.com/not-yet-awesome-rust/not-yet-awesome-rust/* that lists things that should exist in Rust but don't (yet).

Learn by Reading

Although the state of affairs is constantly improving, finding good Rust reading material beyond the beginner level can still be tricky. Here's a collection of pointers to some of my favorite resources that continue to teach me new things or serve as good references when I have particularly niche or nuanced questions.

First, I recommend looking through the official virtual Rust books linked from *https://www.rust-lang.org/learn/.* Some, like the Cargo book, are more reference-like while others, like the Embedded book, are more guide-like, but they're all deep sources of solid technical information about their respective topics. *The Rustonomicon (https://doc.rust-lang.org/nomicon/),* in particular, is a lifesaver when you're writing unsafe code.

Two more books that are worth checking out are the *Guide to rustc Development (https://rustc-dev-guide.rust-lang.org/)* and the *Standard Library Developers Guide (https://std-dev-guide.rust-lang.org/).* These are fantastic resources if you're curious about how the Rust compiler does what it does or how the standard library is designed, or if you want some pointers before you try your hand at contributing to Rust itself. The official Rust guidelines are also a treasure trove of information; I've already mentioned the *Rust API Guidelines (https://rust-lang.github.io/api-guidelines/)* in the book, but a *Rust Unsafe Code Guidelines Reference* is also available (*https://rust-lang.github .io/unsafe-code-guidelines/*), and by the time you read this book there may be more.

NOTE *One of the resources listed at* https://www.rust-lang.org/learn/ *is the Rust Reference, which is essentially a full specification for the Rust language. While parts of it are quite dry, like the exact grammar used for parsing or basics about the in-memory representations of the primitive types, some of it is fascinating reading, like the section on type layout and the enumeration of behavior considered undefined.*

There are also a number of unofficial virtual Rust books that are enormously valuable collections of experience and knowledge. *The Little Book of Rust Macros (https://veykril.github.io/tlborm/),* for example, is indispensable if you want to write nontrivial declarative macros, and *The Rust Performance Book (https://nnethercote.github.io/perf-book/)* is filled with tips and tricks for improving the performance of Rust code both at the micro and the macro level. Other great resources include the *Rust Fuzz Book (https:// rust-fuzz.github.io/book/),* which explores fuzz testing in more detail, and the *Rust Cookbook* (*https://rust-lang-nursery.github.io/rust-cookbook/*), which suggests idiomatic solutions to common programming tasks. There's even a resource for finding more books, *The Little Book of Rust Books* (*https://lborb. github.io/book/unofficial.html*)!

If you prefer more hands-on reading, the Tokio project has published *mini-redis* (*https://github.com/tokio-rs/mini-redis/*), an incomplete but idiomatic implementation of a Redis client and server that's extremely well documented and specifically written to serve as a guide to writing asynchronous code. If you're more of a data structures person, *Learn Rust with Entirely Too Many Linked Lists* (*https://rust-unofficial.github.io/too-many-lists/*) is an enlightening and fun read that gets into lots of gnarly details about ownership and references. If you're looking for something closer to the hardware, Philipp Oppermann's *Writing an OS in Rust* (*https://os.phil-opp.com/*) goes through the whole operating system stack in great detail while teaching you good Rust patterns in the process. I also highly recommend Amos's collection of articles (*https://fasterthanli.me/tags/rust/*) if you want a wide sampling of interesting deep dives written in a conversational style.

When you feel more confident in your Rust abilities and need more of a quick reference than a long tutorial, I've found the *Rust Language Cheat Sheet* (*https://cheats.rs/*) great for looking things up quickly. It also provides very nice visual explanations for most topics, so even if you're looking up something you're not intimately familiar with already, the explanations are pretty approachable.

And finally, if you want to put all of your Rust understanding to the test, go give David Tolnay's *Rust Quiz* (*https://dtolnay.github.io/rust-quiz/*) a try. There are some real mind-benders in there, but each question comes with a thorough explanation of what's going on, so even if you get one wrong, you'll have learned from the experience!

Learn by Teaching

My experience has been that the best way to learn something well and thoroughly, by far, is to try to teach it to others. I have learned an enormous amount from writing this book, and I learn new things every time I make a new Rust video or podcast episode. So, I wholeheartedly recommend that you try your hand at teaching others about some of the things you've learned from reading this book or that you learn from here on out. It can take whatever form you prefer: in person, writing a blog post, tweeting, making a video or podcast, or giving a talk. The important thing is that you try to convey your newfound knowledge in your own words to someone who doesn't already understand the topic—in doing so, you also give back to the community so that the next you that comes along has a slightly easier time getting up to speed. Teaching is a humbling and deeply educational experience, and I cannot recommend it highly enough.

NOTE *Whether you're looking to teach or be taught, make sure to visit Awesome Rust Mentors (*https://rustbeginners.github.io/awesome-rust-mentors/*).*

Summary

In this chapter, we've covered Rust beyond what exists in your local workspace. We surveyed useful tools, libraries, and Rust features; looked at how to stay up to date as the ecosystem continues to evolve; and then discussed how you can get your hands dirty and contribute back to the ecosystem yourself. Finally, we discussed where you can go next to continue your Rust journey now that this book has reached its end. And with that, there's little more to do than to declare:

```
}
```

INDEX

Rust for Rustaceans is set in New Baskerville, Futura, Dogma, and TheSansMono Condensed. The book was printed and bound by Sheridan Books, Inc. in Chelsea, Michigan.

Never before has the world relied so heavily on the Internet to stay connected and informed. That makes the Electronic Frontier Foundation's mission—to ensure that technology supports freedom, justice, and innovation for all people—more urgent than ever.

For over 30 years, EFF has fought for tech users through activism, in the courts, and by developing software to overcome obstacles to your privacy, security, and free expression. This dedication empowers all of us through darkness. With your help we can navigate toward a brighter digital future.

RESOURCES

Visit *https://nostarch.com/rust-rustaceans/* for errata and more information.

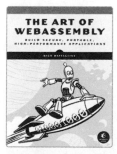